PERFORMANCE FLIGHT TESTING

BY HUBERT "SKIP" SMITH

To my mother and father, who made my career possible, and to the little town of Dillsburg, where I grew up.

FIRST EDITION

FIRST PRINTING

Printed in the United States of America

Library of Congress Cataloging in Publication Data

Smith, Hubert.
Performance flight testing.

Includes index.
1. Airplanes—Flight testing. I. Title.
TL671.7.S52 629.134'53 82-5818
ISBN 0-8306-2340-X (pbk.) AACR2

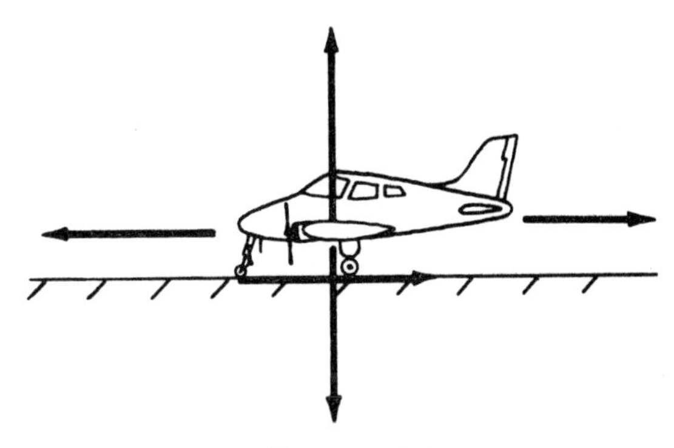

Contents

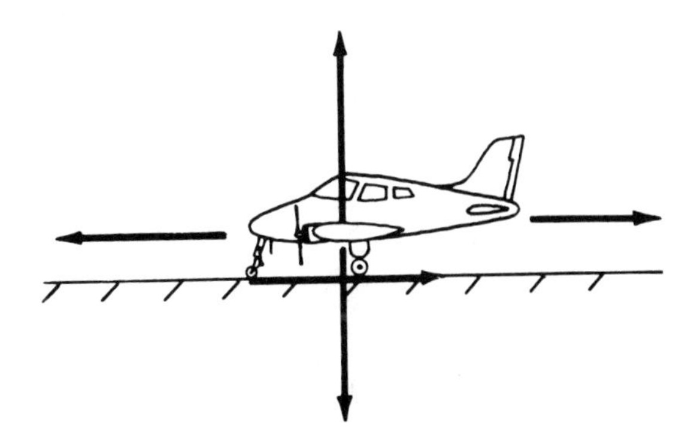

Acknowledgments

I am deeply grateful to Piper Aircraft Corporation and Cessna Aircraft Company for their cooperation in providing performance information on their airplanes, as well as Jeppesen-Sanderson for use of material on their computers. I also thank my wife, Esther, for her help as both model and photographer as well as for her patience while I completed this work; Karen Sweeley, my faithful typist; and John B. Johns and Ed McGarvey for their excellent drawings.

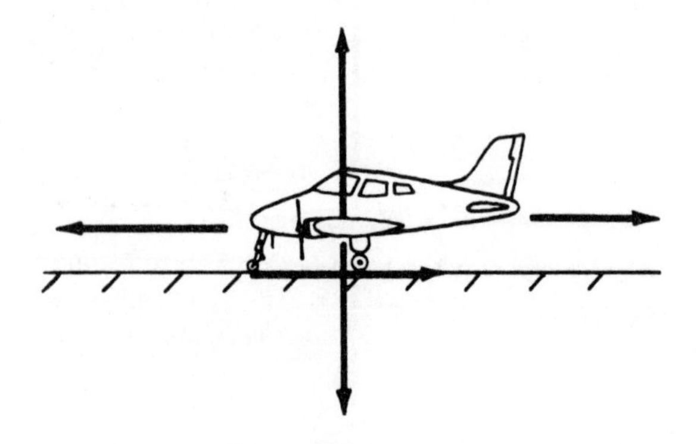

Introduction

In the late 1930s, Clark Gable starred in the movie *Test Pilot*. He was dashing, daring, adventurous, rugged, and not too terribly bright. In appropriate garb he could equally well have played a cowboy, private eye, explorer, sea captain, or any one of dozens of heroes typical of that era. Of course, the helmet and goggles, the leather jacket and boots, and the white scarf unmistakably identified him as a pilot. In this role he threw airplanes around the sky, put them into screaming dives, and otherwise pushed them to and beyond their design limits, often to the chagrin of his employers. Sometimes he barely escaped by parachute as the airplane literally fell apart around him. Thus was established the Hollywood image of test flying.

Real-life flight testing (like most subjects exploited by the film industry) is quite different, however. New airplane designs are tested very carefully and methodically. They are flown only by the most skillful and experienced of pilots whose brains usually exceed their brawn. Their judgment must be honed to an even finer edge than their skill, and procedures must be followed to the letter.

In the early stages of flight testing, the airplane is very gently maneuvered to determine its degree of stability and control. Only after its handling qualities have been determined as acceptable is it pushed to test its endurance near design limits. If any problems arise, it goes back to the drawing board (or computer display, nowadays).

Once the airplane is proved to perform safely, it then progresses to another stage of flight testing. In this stage the airplane is carefully tested to see if it meets the performance that was expected of it when it was designed. This procedure is, therefore, appropriately termed, "performance flight testing." Aircraft manufacturers employ rather sophisticated and complex equipment to make this determination, particularly in the case of military or other high-performance aircraft. For airplanes on the other end of the performance scale, there are many old techniques that are much simpler and work almost as well. Many of these can be employed by the average pilot with little technical experience beyond that normally required to obtain flight ratings. These procedures are pretty much devoid of thrills, involving maneuvers less hair-raising than those typically required to satisfy a commercial pilot examiner. They do, on the other hand, require some precision flying if fairly accurate results are to be obtained. Such precision would need to be at about the level expected in normal instrument flying.

But why go through a lengthy procedure to flight test an airplane that has already been tested by the factory? There are several answers to this obvious question. First of all, the manufacturer tests a brand new, tightly fitted and carefully rigged airplane. Aircraft deterioriate, aerodynamically, with age. Scratches and dents gradually appear; door, window, and other

The performance capability of homebuilt aircraft is pretty much unknown until proven by flight tests.

Older aircraft, such as this 1953 Bonanza, are also good candidates for performance re-evaluation, due to aerodynamic deterioration over the years.

seals begin to leak; patches are added to repair minor damage; and the continual disassembly and assembly required for inspections cause minor distortion and misalignment of airframe components. None of these flaws means very much in itself, but collectively they can add up to a significant decrease in performance. There is also decreased power output with engine wear. New airplanes also have brand new engines when performance is measured. All of these factors can mean a noticeable difference between the performance of an airplane right off the production line and one 15 years old with 5000 hours on it.

Then, too, not all airplanes are exactly alike, even when new. Surely you have seen ads for used airplanes where the craft was extolled as "exceptionally fast for a Skybird II." Some of these claims may not be exaggerations. Certain airplanes just turn out to be slightly better performers than their more normal siblings. By the same token, some also turn out to be a bit more doggy. You never see *this* fact mentioned in ads, however.

So far, we have been discussing production airplanes. If you are completing a homebuilt, then the necessity for flight testing should be obvious. Even those built to the exact specifications of the designer will vary considerably with construction technique. And what homebuilder has not substituted here and there for costly or unavailable materials or parts? In fact, one of the big attractions of homebuilding is the freedom to customize the airplane. You usually have a choice of engines, propellers, wheels, instruments, and other accessories. The temptation to make further modifications is great. Changing from conventional to tricycle landing gear or adding wheel fairings can often be accomplished without completely redoing the original design. But these modifications cause changes in performance. Even the addition of a canopy to an open cockpit can make a difference. Consequently, very few homebuilts of the same basic design are really alike. If you fly one of these, you really should test it to determine its particular performance. Of course, this goes without saying if you are building an original design.

There is one more reason to go through a series of performance flight tests and that is simply to learn more about airplane performance. Certainly, you can refer to the charts in the operator's handbook if you fly a fairly average production airplane. But the numbers here take on a deeper meaning when you have run through the tests and verified them for yourself. There are also a number of things revealed that do not show up in standard performance charts. Few pilots know such things as the rate-of-climb when flying at other than best rate-of-climb speed or the L/D ratio at various airspeeds. Even the maximum L/D ratio is an obscure number to veteran throttle pushers. The variation in stall speed with gross weight or the change in takeoff distance with density altitude will become very apparent as you plot it out yourself. In short, testing your airplane will enable you to understand its performance a lot better and will, consequently, make you a safer pilot.

This book describes a series of tests by which most of the performance information normally required for airplane operation can be obtained. The tests are designed to be carried out by the average light airplane pilot or someone with equal aviation background. No special equipment is required other than a stopwatch and a basic electronic calculator. In some cases, such items as a tape recorder or camera are helpful but not absolutely essential. The standard instruments in the airplane are the primary devices used to obtain the necessary measurements. For many of the tests only an altimeter, an airspeed indicator, and an outside air temperature gauge are required.

The first two chapters deal with background material and general procedures to be followed in conducting the tests and obtaining and reducing flight data. The next chapter explains the calibration of the airspeed indicator so that accurate information on airspeed is available for the subsequent tests. Chapter 4 outlines tests for the determination of stall speed, which is also needed before some of the other tests can be performed. The next four chapters deal with the primary performance areas of takeoff, climb, cruise, and descent and landing, respectively. In addition to describing the flight test procedure for each of these areas, some discussion is included to explain the reasoning behind the procedure and why the airplane performs as it does. This information is presented in a very basic way and does not involve complex mathematics or require any special technical background for understanding of the material.

An appendix is included which lists performance data of some actual airplanes as presented in the aircraft operating handbooks. This information is provided as a guide to show how performance data is normally presented. Upon completion of all of the tests outlined, you should be able to construct a similar set of charts custom fitted to your own individual airplane. Users of such data are cautioned to interpret it with a degree of wisdom. All of the tests are based on sound scientific principles. However, many approximations are used which do not yield *exact* results. It would be rather foolish, for example, to attempt a takeoff on an 1800-foot runway even though the performance you measured indicates that 1790 feet are required. Use some judgment and adequate margins of safety. Neither I nor the publisher accept responsibility for the accuracy of the data obtained from tests described herein. Piper and Cessna will not be liable for information in their respective sections of the appendix, since this information is not kept current.

Chapter 1

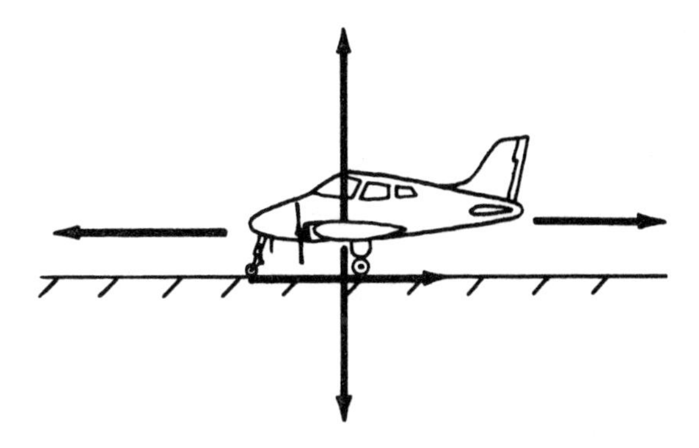

Effects of the Atmosphere

In addition to the characteristics of the airplane itself, performance depends a great deal on the nature of the atmosphere through which the airplane flies. Unfortunately, the atmosphere varies quite a bit in its makeup from day to day, and also, with altitude on any given day.

The atmosphere is a big mass of air which covers the earth like an ocean covers the ocean floor. Unlike the sea, however, the atmosphere gets less dense, or lighter in weight, as altitude is increased. This characteristic is due to the fact that air is compressible fluid (water is not). The air is held against the earth by the attraction of gravity, which means that it has a certain amount of weight to it, and it is compressed by this weight. It is compressed most at the earth's surface because all of the atmosphere is above that level and all of it is bearing down on the layer right at the surface. As we go higher, say to 50,000 feet, only the atmosphere *above* 50,000 feet is bearing down on that level. The air up there is compressed less. The same volume of air has less weight at altitude than at the surface because of this lower compression.

The measure of how much weight a certain volume of a substance has is referred to as its *density.* A cubic foot of water weighs 62.4 lbs.; a gallon of gasoline weighs 6 lbs. We call this value of 62.4 pounds per cubic foot or 6 pounds per gallon the density. These fluids, being incompressible, have a fixed density. They can vary *slightly* with temperature, but at normal temperatures are always very close to these weights. Air, on the other hand, is extremely variable. There is no fixed weight per cubic foot associated with it. Air density is very important, however, since all aerodynamic forces, including lift and drag, directly depend on it. The engine power output is also proportional to the air density in which it operates. We must, therefore, know the density of the air in order to determine the airplane's performance.

Density of the air cannot be measured directly. Fortunately, though, density depends on some other properties of the air, namely pressure and temperature, which can be measured. Density is directly proportional to the pressure of the air and inversely proportional to the temperature. This means that the higher the pressure, the higher the density, but the higher the temperature, the lower the density will be.

Mathematically, this relationship is given as:

$$\rho = \frac{P}{T} \times \text{constant}$$

This equation says that ρ, the Greek symbol used for density in scientific work (pronounced "roh") is equal to pressure divided by temperature multiplied by a constant (a number). Thus, once pressure and temperature are known, the density can be calculated.

Pressure and temperature, of course, vary from day to day, from one location to another, and with altitude. However, there are certain average values of these quantities which exist at the earth's surface. Also, since the altitude of the surface varies, there is a standard reference base established from which altitude is measured. This reference is the average height of the sea and referred to as "mean sea level" or MSL. The average conditions established as standard at mean sea level are:

P = Pressure = 29.92 in Hg (inches of mercury)
T = Temperature = 59°F

These two values will also give a standard density when used in the equation given above of:

$$\rho = \text{Density} = 0.0765 \text{ lbs/cu. ft.}$$

Density is also sometimes measured in terms of *mass* per cubic foot. Mass is weight divided by the gravitational constant and has units called *slugs*. Standard sea level density in these units is 0.002378 slugs per cubic foot. In addition to these standard values at sea level, the standard rate of temperature change with altitude is −3.579° per thousand feet. This is sometimes referred to as the "standard lapse rate."

Using the standard value of temperature at MSL and the standard lapse rate, standard values of pressure and density can also be determined for various altitudes. On a *standard* day these values would really exist at those actual altitudes. Table 1-1 shows the standard values of temperature, pressure, and weight density at altitudes up to 20,000 feet. On a standard day, for example, at 6000 feet the temperature would be 37.6°, the pressure 23.98 in. Hg, and the density 0.0640 lbs/cu. ft.

On most days, however, the atmospheric properties are not standard. The pressure at 6000 feet MSL may not be 23.98, and the pressure at sea level may not be 29.92. However, there will be *some* altitude at which the pressure is 29.92 (standard sea level pressure), and 6000 feet above this, the pressure will be 23.98. This altitude is referred to as 6000 feet *pressure* altitude, regardless of the true altitude at which it is located. Pressure altitude is the height of a respective value of pressure above the height where 29.92 in. Hg is located. Remember that on a *standard* day this is the value right at sea level.

Table 1-1. Standard Atmospheric Properties.

Altitude (ft)	Temperature (°F)	Pressure (in Hg.)	Density lbs/cu. ft.
0	59.0	29.92	.07651
1,000	55.4	28.86	.07430
2,000	51.9	27.82	.07213
3,000	48.3	26.82	.07001
4,000	44.7	25.84	.06794
5,000	41.2	24.89	.06592
6,000	37.6	23.98	.06395
7,000	34.0	23.09	.06202
8,000	30.5	22.22	.06013
9,000	26.9	21.38	.05829
10,000	23.3	20.57	.05649
11,000	19.8	19.79	.05474
12,000	16.2	19.02	.05302
13,000	12.6	18.29	.05135
14,000	9.1	17.57	.04972
15,000	5.5	16.88	.04813
16,000	1.9	16.21	.04658
17,000	−1.6	15.56	.04507
18,000	−5.2	14.94	.04359
19,000	−8.8	14.33	.04215
20,000	−12.3	13.74	.04075

To illustrate pressure altitude, consider the airplane in Fig. 1-1. On a standard day, the pressure is 29.92 at sea level. If the pilot correctly sets this value into his altimeter it will read zero at sea level and 6000 feet at 6000 feet true altitude (assuming no instrument error). Now let us assume that the pilot flies into a lower pressure area. The pressure at sea level here is 28.86. If he resets his altimeter to this value he will continue to read 6000 feet at that altitude and if he were to descend to sea level the altimeter would read zero. Notice now, though, that the *standard* sea level pressure of 29.92 doesn't exist at sea level but 1000 feet *below* sea level. Flying at 6000 feet in this area, the airplane is actually 7000 feet above the baseline altitude where the pressure is 29.92. We, therefore, say that the airplane, while at 6000 feet *true* altitude, is at 7000 feet *pressure altitude*. The pilot could easily determine this altitude by resetting the altimeter to 29.92, in which case the altimeter would now read 7000 feet. Therefore, setting the altimeter to the actual sea level pressure gives an altitude reading in *actual*, or true, altitude; setting it to 29.92 gives a reading in *pressure* altitude.

This same sort of reasoning applies to another type of altitude which is called *density altitude*. If the airplane were at 6000 feet on a standard day, it would be 6000 feet above sea level and also the level where the density is 0.0765 lbs/cu. ft. Again, however, let us

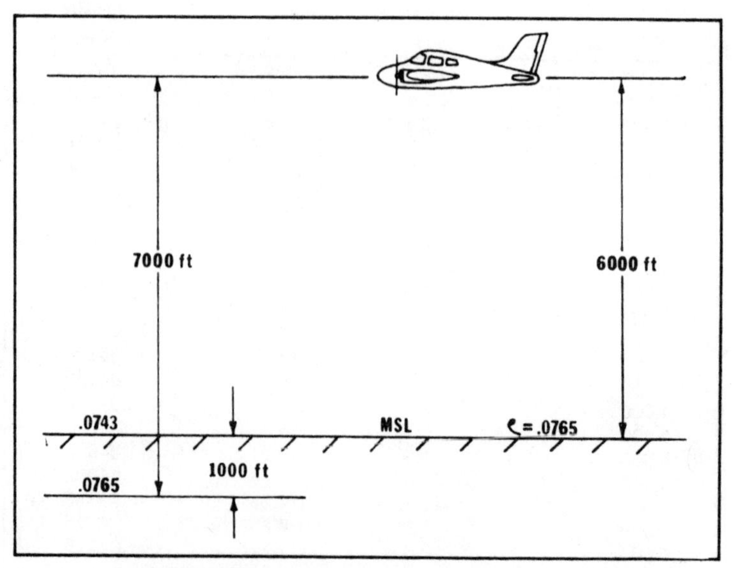

Fig. 1-1. Pressure altitude relationship showing the height of the airplane above the level where the pressure is 29.92 in. Hg.

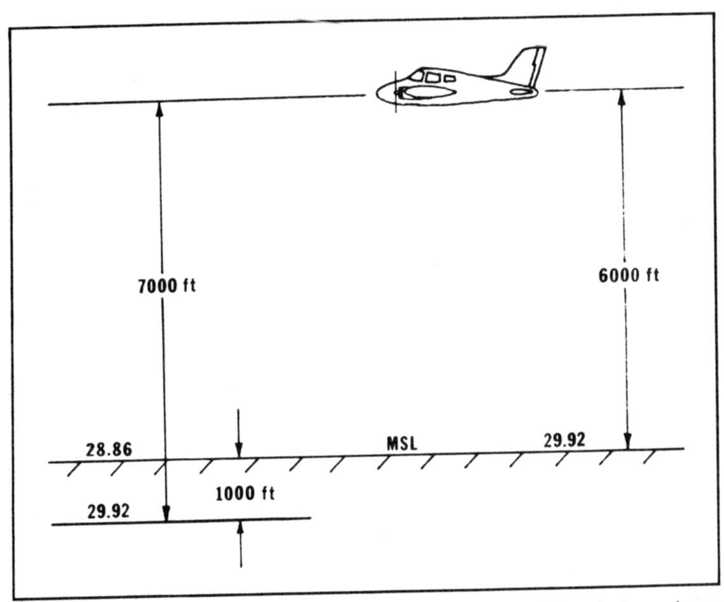

Fig. 1-2. Density altitude relationship showing the height of the airplane above the level where the density is 0.0765 lbs /cu. ft.

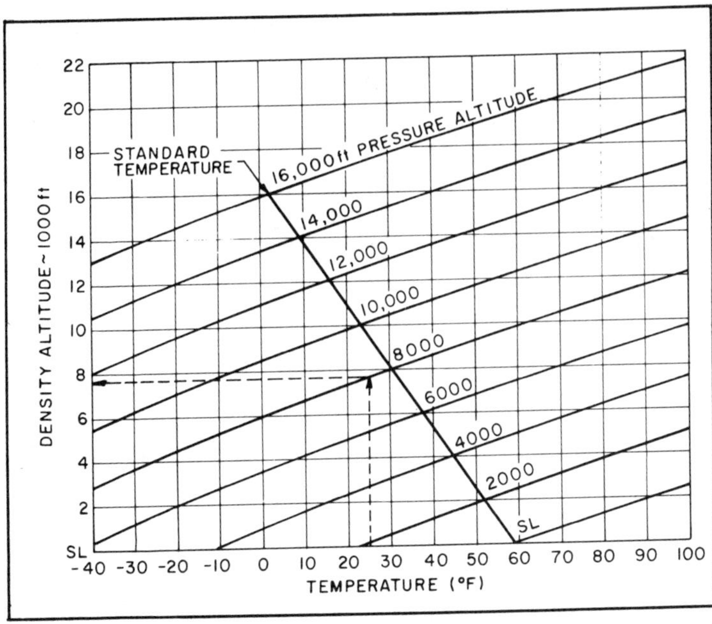

Fig. 1-3. Chart for the determination of density altitude from pressure altitude and temperature.

consider a nonstandard situation. Suppose the combination of pressure and temperature at sea level resulted in a density of 0.0743 lbs/cu. ft. This is the density that occurs on a standard day at 1000 feet. In order to reach 0.0765, one would have to descend 1000 feet below sea level. Therefore, in this situation, at a true altitude of 6000 feet, the airplane is flying in density that normally occurs at 7000 feet as shown in Fig. 1-2. This altitude is thus referred to as 7000 feet *density altitude*.

Just as the actual density can be calculated from the pressure and temperature, density altitude can be calculated if the pressure altitude and temperature are known. We have already discussed the way in which pressure altitude can be determined from the altimeter. Finding the temperature is even easier. All that one needs to do is to read the outside air temperature gauge, or OAT, as it is called. The simplest way, then, to proceed to determine density altitude is to enter the appropriate temperature opposite the appropriate

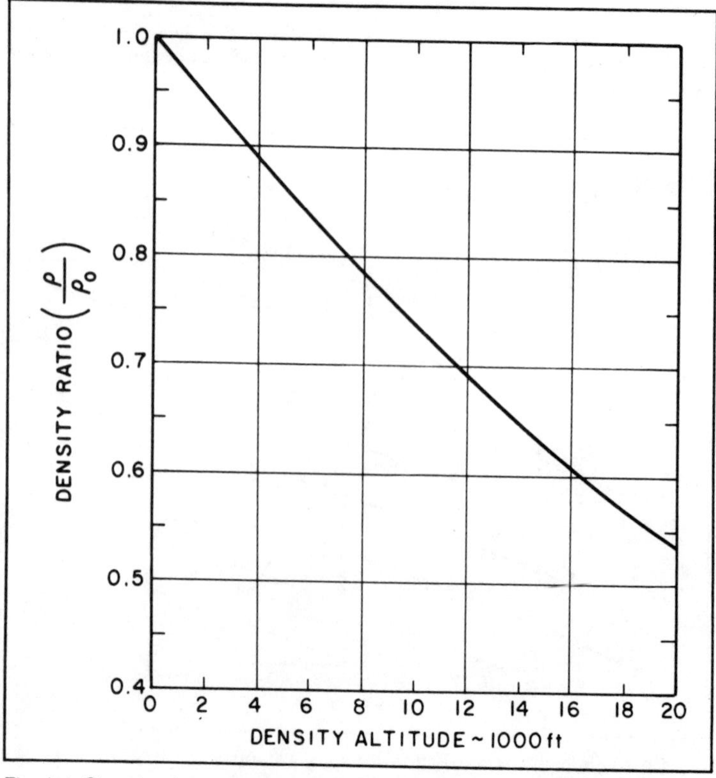

Fig. 1-4. Chart for determining density ratio for the appropriate density altitude.

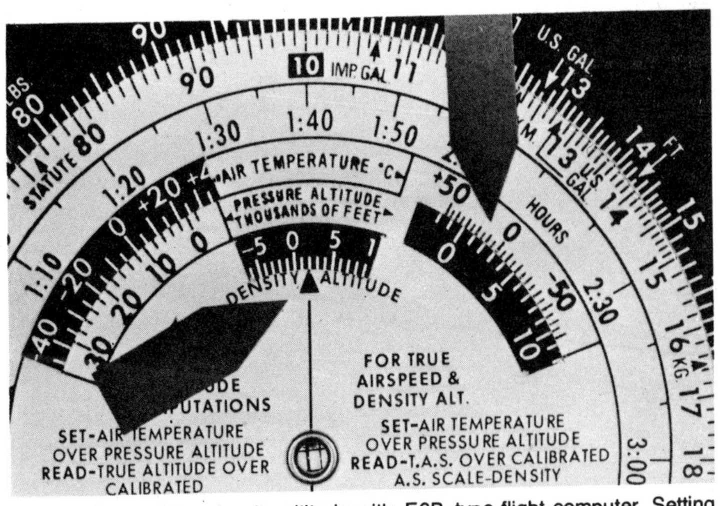

Fig. 1-5. Determining density altitude with E6B type flight computer. Setting temperature opposite pressure altitude yields density altitude in the density altitude window.

pressure altitude on the flight computer. The density altitude can then be read in the density altitude window of the computer.

Determining density altitude in this way with an E6B type computer is rather inaccurate due to the very small graduations on the density altitude scale. A better way is to use the chart shown in Fig. 1-3. This chart is typical of many included in airplane information manuals. Pressure altitude and temperature must first be determined, as before. Then the chart is entered along the bottom with the appropriate temperature. The dotted line shows an example using 25°F. Go up vertically until reaching the line for the appropriate pressure altitude (in the example, 8000 feet). Then read directly to the left on the vertical scale, the density altitude. In the example, this is about 7500 feet. Be sure to use the proper temperature scale. The chart is plotted in degrees Fahrenheit, while most computers use degrees Celsius. An even better way of finding density altitude is with an electronic flight computer, which is further discussed in Chapter 2.

Occasionally, the actual density, or sometimes density ratio, is required for a given altitude. The density ratio can be obtained from Fig. 1-4 once the density altitude is determined by one of the methods outlined above (Fig. 1-5). The density ratio at 14,000 feet, for example, is 0.65. This means that the density here is 65 percent of what it is at standard sea level (or zero density altitude). The density, then, would be 65 percent of 0.0765 or 0.0497 lbs/cu. ft.

Chapter 2

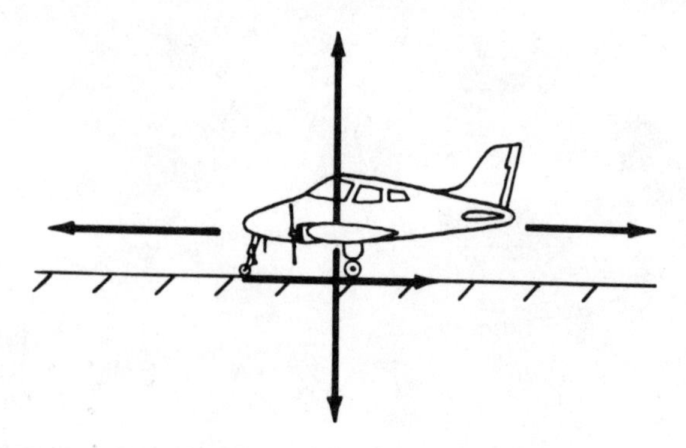

General Test Flight Procedures

As with any flight operations, the flight tests described in this book must be conducted within the confines of the Federal Aviation Regulations and the operating limitations of the airplane. None of the tests require exceeding either of these restrictions. However, it is essential that the pilot observe all of the rules. The airspeed calibration test, for example, demands a low altitude, but FAR 91.79 regarding minimum safe altitudes should be followed. Likewise, a safe terrain clearance altitude should be attained before performing tests such as stall speed. It would be wise to observe FAR 91.71 dealing with aerobatic flight for all altitude flight tests, even though many do not necessarily involve abrupt altitude change maneuvers.

All test flying should be performed in good VFR weather; that is, not just marginally VFR. The tests should be conducted in non-congested airspace. Even so, it is particularly important to be alert for other traffic. The nature of test flying requires a concentration of the pilot on the instrument panel in order to hold exact conditions, much like IFR flight. The observer(s) will also be concentrating on reading instruments, recording data, or spotting landmarks. It is a very good idea to also carry one more crewmember than necessary for carrying out the tests, if space permits. This person can serve as a traffic spotter and should be someone familiar with this sort of activity.

If some crewmembers are not pilots or not accustomed to flying in light aircraft, they should be thoroughly briefed on safety and emergency procedures. Make sure that they understand how and when to use seatbelts. It is also wise to brief them on the meaning of warning horns, such as stall and gear warnings which may sound, or other unusual situations that may occur during the course of conducting some tests. This helps to relieve much anxiety in such situations and makes the testing easier and more pleasant. A calm data-taker also records much more accurate information than a nervous one.

Methods of Obtaining Data

More than one crewmember is required for most tests. Make sure that the duties of each are understood prior to takeoff (Fig. 2-1). The pilot needs to concentrate on accurately flying the airplane. A second crewmember may need to continually observe instrument indications. A third person is often desirable to record data read out by the observer. In other tests, landmark spotting is required by one person. Another may serve as both instrument observer and data recorder in such situations. Usually, three people are required to comfortably perform the tests. Most tests could possibly be performed, however, by two and would have to be in the case of a two-place airplane.

Data is obtained, in most cases, from the standard instruments installed in the panel. If the observer is not a pilot, make sure that he

Fig. 2-1. Planning and coordinating duties among crew members prior to the flight test.

Fig. 2-2. Recording data on a formal data sheet. Such a sheet is given in Appendix A.

or she is familiar with the instrument indications. Also, make sure that everyone in the crew understands what data will be taken and in what order. It would be wise to rule off columns on a sheet of paper to be used as a data sheet. Label the column headings for the data to be taken in the order in which it will be read. A well coordinated effort among pilot, observer, and recorder will make testing much easier and avoid confusion later when the data are reduced and plotted. Appendix A contains a typical data sheet (Fig. 2-2).

Practically all tests will require an airspeed value. Usually, the true airspeed is ultimately required. However, the value as read from the airspeed indicator is the indicated airspeed (IAS). It is the value that you must start with in order to calculate a true airspeed. Methods for correcting to calibrated and true airspeed are discussed in Chapter 3. Some airspeed indicators are calibrated in knots and some in miles per hour (Fig. 2-3). Other indicators have both scales. It doesn't really matter which scale is used as long as you are consistent. If both scales are indicated, it is probably best to read statute mph simply because the graduations are larger and, hence, the reading will be more accurate. If airspeed information is desired in knots, mph readings could be converted later.

Performance of the airplane is also highly dependent on the density of the atmosphere as discussed in Chapter 1. Since density can't be measured directly, we depend on measuring temperature and pressure in order to calculate density. Pressure is actually measured in terms of pressure altitude. The altimeter is used for this purpose and indicates in *pressure* altitude whenever the barometric pressure in the altimeter window is set to 29.92 inches of mercury. Since pressure altitude (not true altitude) is what is required in all flight tests, the altimeter should always be set at 29.92 during the testing (Fig. 2-4). Takeoff, landing, and other flight operations in getting to and from the test area, however, require a knowledge of *true* altitude. Therefore, it is best to set the altimeter

Fig. 2-3. Typical airspeed indicator. When two scales are given, the MPH scale is easier to use because it is larger.

Fig. 2-4. Altimeter showing pressure altitude when set to 29.92 in. Hg.

to the proper barometric pressure at the time and location (or field elevation) prior to take-off. After takeoff and climb to the test location, the altimeter can be reset to 29.92. It is, also, a good practice to record the true pressure so that the altimeter can be reset to true altitude for descent and landing. Read the altimeter and record the data in feet of pressure altitude. Since the smallest increment on most altimeters is 20 feet, the nearest 20 (or possibly 10 feet) of altitude is sufficient for accuracy. Ten feet or so of altitude makes very little difference in density.

Temperature is obtained from the outside air temperature gauge, or OAT. Most of these gauges also have two scales, one in degrees Celsius and one in degrees Fahrenheit (Fig. 2-5). It is again more convenient to read the larger scale, which is the Fahrenheit scale. The charts and calculations in this book are thus based on temperature reading in °F. Read the temperature as closely as possible to single degrees, for example, 54° or 57°, even though the smallest increments on the scale are 5 degrees.

For fuel consumption determination in calculating range, the fuel flow meter can be utilized if one is installed in the panel. Such meters are usually standard on airplanes with fuel-injected engines. This gauge reads either in gallons or pounds of fuel per hour. Although not required to calculate the performance, it is necessary to know the proper power settings for a certain percentage of rated power in the cruise performance tests. Power is set by the appropriate rpm with a fixed-pitch propeller or combination of rpm and manifold pressure with a constant-speed propeller. Therefore, at-

Fig. 2-5. Typical outside air temperature gauge. The outer, or Fahrenheit, scale is larger and, thus, more easily read.

tention should be given to the tachometer and manifold pressure gauge to ascertain correct power settings for cruise speed and range determination (Fig. 2-6).

To summarize, the following instruments will be utilized to obtain the respective information:

- ☐ Airspeed indicator—indicated airspeed.
- ☐ Altimeter—pressure altitude.
- ☐ Outside air temperature gauge—temperature.
- ☐ Tachometer and MAP gauge—engine power.
- ☐ Fuel flow meter—fuel consumption rate.

Fig. 2-6. The tachometer and manifold pressure gauge, used for setting specific amount of engine power. Also included is the fuel flow meter, used to determine fuel consumption.

These instruments are shown in Figs. 2-3 through 2-6. For many tests, only the first three of these instruments will be required to yield the necessary information for measuring performance. In some tests some timing must be done and is best performed with a hand-held stopwatch. A portable tape recorder and a camera are also useful for recording data in certain tests. However, these devices are not essential. You may want to use your ingenuity and come up with some original devices or applications of devices for variations in some of the testing techniques.

In the event that you want to test a single-place airplane, the procedure is a bit different. Obviously, the pilot is the whole crew. He will have to fly the airplane, do the observing, and record the data. Needless to say, he will have his hands full. However, this feat is not impossible. If you are the pilot (and crew) of a single-place, it would be desirable to use a tape recorder for recording much or all, of the data. This procedure eliminates the distraction from observation of the instruments while writing down the readings. A stopwatch mounted on the panel or yoke, at least temporarily, would also be a big help. Develop your own sequence of reading and recording information which seems to be most logical and convenient for you. Don't forget that you are also the traffic observer in this case. Pay particular attention to choosing a safe area and time for testing. It would be wise to also make frequency clearing turns between runs or steps in the test flights.

Gross Weight Determination

The weight of the airplane has an effect to some degree on all items of performance. The effect is quite significant on certain performance such as takeoff, climb, and stall speed. For these tests, therefore, it is important to know the weight rather accurately. In other tests, such as maximum speed and range determination, the weight has a lesser effect and some variation in the weight can be tolerated.

Total weight, of course, means the entire weight of the airplane as flown, including fuel and payload at any particular time. Total weight is sometimes referred to as *gross* weight. There are maximum gross weight limitations on each airplane type. Sometimes these limits are different for takeoff and landing. For most single-engine airplanes, though, they are usually the same. In this case the value is referred to simply as the maximum gross weight.

Some performance, such as climb and cruise, is usually presented in operator's manuals for one gross weight. This weight is known as the *standard* weight and for light single-engine airplanes is normally the maximum gross weight. Other performance, such as takeoff and landing distance, is presented for a variety of weights. For the tests in this book, weight corrections can be made, in some cases for weights other than standard. This is true for stall speed and takeoff and landing performance. For others, it is necessary to establish a standard weight and make all tests as close to that weight as possible.

The question now arises as to what is standard weight. The maximum allowable gross weight could be chosen for this value. Another choice, and perhaps a more practical one, is a normal operating weight. Such "normal" weight would depend on your particular flying habits. If you *normally* carry four people and baggage with full fuel in your four-place airplane, then the maximum gross weight is pretty normal. On the other hand, if typical flights usually involve only two people without baggage in a four-place craft, then the average weight in such configuration might be a more normal value for you. The advantage of using maximum gross weight is that this value would yield the minimum performance and any lower weight would improve upon it. In this way your performance is always on the conservative side. Rate-of-climb, for example, at lower weight would be greater than that at maximum gross and takeoff distances would be shorter. The choice is up to you, but then you must adhere to that weight in the loading of the airplane during the course of the flight tests.

For all tests it is desirable to know the weight of the airplane fairly accurately. The best way to determine this is to fill the fuel tanks completely before each flight test. In this way there is no doubt about the exact amount of fuel aboard. Then record the weight of each crew member. If any of the crew are uncertain of their weights, have them weighed. Remember to take the weight as dressed for the flight. With heavy winter clothing this weight could vary significantly from that taken in a doctor's office. Add the weight of the full fuel and crew members to the empty weight of the airplane. If the airplane is experimental or modified from the configuration when it was originally weighed, an actual weighing of the airplane must be performed to determine empty weight. If your selected standard weight is higher than the total of the empty weight, fuel weight, and crew weight, you can either carry additional passengers or some sort of ballast. Lead weights, sandbags, or even bundles of magazines or newspapers will suffice. Remember to tie them down and place them so that the CG is within limits.

During the flight, fuel will be consumed and the weight will be reduced somewhat. After landing, refuel the airplane and note the mount of fuel consumed (Fig. 2-7). Multiply the gallons of fuel used by 6 to get fuel in pounds. Then take half of this value and subtract it from your takeoff weight to get an *average* value of gross weight for the tests. Let's consider an example.

Suppose your airplane has an empty weight of 1400 lbs and a maximum gross weight of 2500 lbs. You have decided that you normally operate at about 2400 lbs and that you want to evaluate performance at this weight. The fuel capacity is 50 gallons usable and you require a crew of three whose combined weight is 510 lbs. Your gross weight would be computed as follows:

Empty weight:	1400 lbs
Fuel weight (50 gal.):	300
Crew weight:	510
Total	2210 lbs

To bring this figure up to your standard weight of 2400 lbs, you add a passenger who weighs 160 lbs and 30 lbs of ballast.

Now, let's assume that you flew for one hour and on refueling the airplane required 10 gallons of fuel or 60 lbs. The average operating weight would be 2400 lbs minus half of 60 or 2370 lbs:

$$W_{av} = 2400 - \frac{60}{2} = 2370 \text{ lbs.}$$

This figure would then be the exact weight to use in your calculations. If you knew that the fuel consumption was about 10 gph and

Fig. 2-7. Monitoring refueling operations after a flight. Actual fuel consumption can be used to determine an average weight during the flight tests.

that your flight would take about an hour, you could have loaded to 2430 so that the average weight would come out to right about 2400 lbs. Thirty pounds, however, is only slightly over one percent of 2400 pounds and should make little difference in performance. Of course, you could choose the maximum gross weight of 2500 lbs as standard and load the airplane to that weight to begin your tests. Since you should not ever exceed the maximum gross weight, your average operating weight would always have to be somewhat less than this value.

Weather for Flight Testing

All pilots are aware of the fact that weather conditions impose severe restrictions on flying. Adequate ceilings and visibility must exist throughout the entire flight if it is to be made under VFR conditions. Even IFR operations are not possible if clouds are too low at the destination or if such demons as thunderstorms and icing conditions are present. Flight testing is, probably, the flight operation most sensitive to weather conditions. Not only must the weather be well above VFR minimums for adequate visibility, but it must also be free of turbulence. Turbulence, or the presence of vertical currents in the air, is the biggest obstacle to obtaining good test data.

Turbulence arises from two primary sources. One is the existence of convective currents in the atmosphere. The other is strong wind blowing over uneven terrain. The first of these causes of turbulence, convective currents, can be present even under very calm weather conditions. Convective currents are vertical movements of the air caused by uneven heating of the surface of the earth. Since the heat-reflecting quality of the surface is variable, depending on the nature of the surface material, convective currents vary considerably from one point to another. Green fields and forests reflect relatively little heat, while sandy, dry soil or paved areas reflect considerable heat. For every rising current over these high heating surfaces there is a compensating downward current over the cooler areas. This situation gives rise to vertical motion of the air, both upward and downward. As an airplane flies through these currents, the alternating upward and downward motion is felt as bumpiness to the occupants. The more severe the vertical currents, the rougher the ride will seem.

Good flight test data requires holding the airplane at very precise speeds and attitudes for almost all of the experiments described. Obviously, such procedure is not possible if the air is rough. Furthermore, many tests involve a determination of the airplane's performance in a vertical direction. Rate-of-climb tests and rate-of-descent tests are good examples. If vertical currents are present in the atmosphere, false readings of the airplane's natural climb or descent capability are bound to occur. Sailplane pilots even take advantage of vertical currents to give climb capability to their aircraft, which have absolutely none in calm air. Therefore, certain tests, such as those mentioned, are even more sensitive to atmospheric conditions than others. All tests, however, require relatively smooth air.

Since convective currents, or *thermals*, as they are often referred to by pilots, require heating of the air, they are less severe on cool days. To be more accurate, they are less severe when there is less change in the surface temperature. If the sun comes out very bright and direct after a very cool night, the surface temperature tends to change rapidly and convective currents arise rather rapidly. When the sun is shielded by an overcast, and, in particular, after a fairly warm night, little temperature change occurs to the surface and little thermal activity results. It also takes some time for the surface to heat up, even on bright days. Therefore, early morning is often a good time to run tests. Another good time is just before sunset in the evening, since thermal activity usually subsides at this time. Tests could actually be done at night. However, visual clues and recording of data would be somewhat complicated. Night flying also requires a bit more alertness to factors outside of the airplane. Thus, even though very smooth air can be found at night, it is not recommended as a good time for conducting flight tests.

If testing is done in the morning, it is a good idea to first perform any tests that require low altitudes. This is because the effects of convective currents are lowest early in the morning. As the day progresses and the surface becomes warmer, the turbulence level will rise. The rate-of-climb test, for example, which requires testing at various altitudes, should be started at the lowest altitude. Then, the next higher altitude can be run and so on. In this way you can often keep above the level of thermal activity, even though it does build up.

One way to determine the height of the turbulent air is to note the level of cloud formation. As a convective current rises it cools down due to the cooler air through which it is rising. When it cools to the temperature of the surrounding air, it will not go any higher. If any moisture was present in the rising air, it will condense out and form a cloud when sufficiently cooled. A scattered deck of cumulus

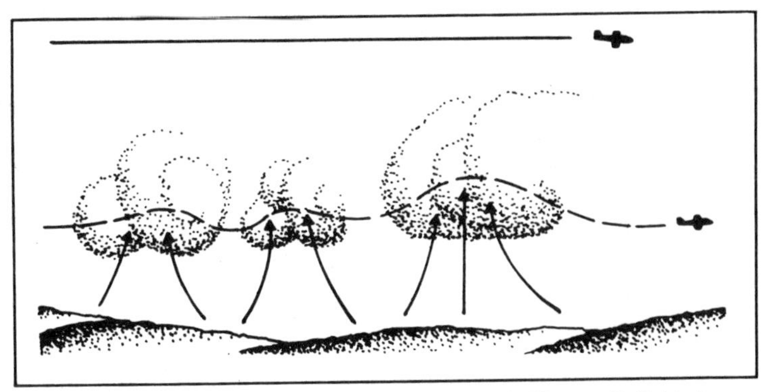

Fig. 2-8. Thermal activity shown ending at cloud level with smooth flight above.

clouds, therefore, is a good indicator of the upper boundary of the turbulent air. When sufficient moisture is not present to form clouds, often dust particles or other fine material is usually carried up by the vertical currents and forms a haze layer. The top of the haze layer is, therefore, another good clue to the location of smooth air (Fig. 2-8).

The other major cause of turbulence is strong wind blowing over mountains or other obstructions (Fig. 2-9). An upward motion is caused on the upwind side of a mountain and a downward motion on the downwind side. In addition, a turbulent wake is formed behind the mountain and rather unpredictable air currents are formed. The effects of such turbulence are much like those due to convective currents. The air is rough and makes the holding of accurate conditions difficult. The upward and downward motion also

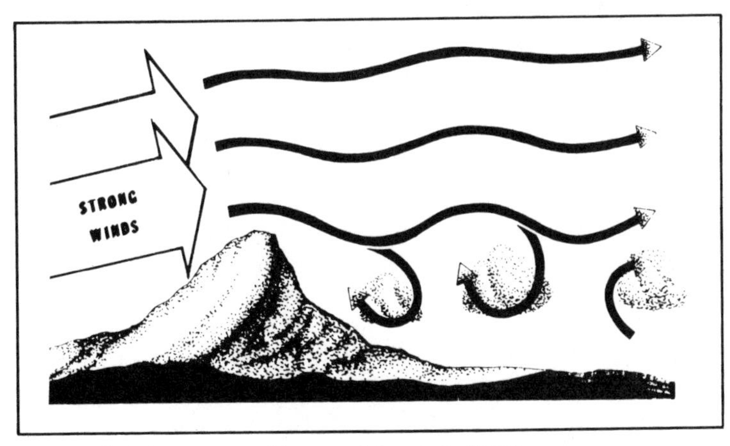

Fig. 2-9. Turbulence caused by mountain wave activity.

imposes unnatural climbing and descent forces on the airplane. Wind-induced turbulence can sometimes be avoided by climbing above it. However, on days when the wind is very strong, the effects may be felt even at very high altitudes. If low altitude tests are required, they may not be possible at any time of the day, if strong winds persist. If you must fly with some amount of wind blowing, make an effort to avoid mountainous or uneven terrain, if at all possible.

Another effect that wind can have is to form into very long waves. Such waves may not impart a very rough ride to aircraft flying through it, since the upward and downward currents are widely spaced. However, strong longwave currents can have significant vertical components to their motion. The airplane could, therefore, be rising or descending due to such wind currents, even though very little roughness in the air is being experienced. It is a good idea to check the consistency of data taken on days when such wind currents could be present. If very unreasonable climbs and descents are being measured, wave effects may be the reason. It is best to cancel the flight tests on such a day and just enjoy the ride.

Use of Equations

A number of mathematical calculations are necessary to determine performance from the tests outlined in this book. Most of these can be done "by hand," which means by pencil and paper and the standard methods of multiplying, dividing, adding, and subtracting that you learned in school. A simple electronic calculator with only the basic functions would be of tremendous help, however. The use of such a device is highly recommended.

One type of calculator that is particularly useful in performing the calculations required in the following tests is the relatively new electronic flight computer. This device, such as the Jeppesen-Sanderson Avstar (Fig. 2-10), electronically performs all of the functions of the E6B type flight computer, in addition to the arithmetic functions of a standard calculator. It is extremely helpful in determining density altitude and true airspeed under various atmospheric conditions. It can also very easily convert knots to miles per hour or degrees Celsius to degrees Fahrenheit and perform many other special aeronautical functions. In addition, however, it can be used as a regular calculator to work out the equations which will be necessary to reduce the test data.

Regardless of the method of calculation, you must have an understanding of the meaning of the equations and the way in which

Fig. 2-10. AVSTAR electronic flight computer (courtesy of Jeppesen-Sanderson, Inc.).

they are set up. If it has been a while since algebra class or, if all of your time there was spent observing the little red-headed girl in front of you, then the following review may be helpful. If, on the other hand, you are a math whiz, then you can skip this section.

Most of us are familiar with the notations for addition and subtraction. $A + B - C$, for example, simply means to add the quantity A to quantity B and subtract quantity C. Multiplication is indicated when two quantities appear side by side, such as AB. This means A multiplied times B. The same meaning is implied if one or both quantities is surrounded by parentheses, such as (A) (B). A divided by B is indicated by placing A over B, as in fractions. A slash means the same thing, such as A/B. Such a term is also sometimes referred to as a *ratio* of A to B.

Higher powers are indicated by superscript numbers. A^2 means A squared or A multiplied times itself. The square root of A is indicated as $\sqrt{A}$ which means a number which, when multiplied by itself, results in A.

Most equations involve combinations of these operations. Brackets and parentheses are used to show the grouping of quantities and the sequence of operations. Consider the following example of an equation to determine a quantity P:

$$P = \left[\frac{A + (B + C)^2}{D}\right]\sqrt{E}$$

This equation says to add B to C, square this sum, add it to A, divide that amount by D, and then multiply the result (everything inside the brackets) by the square root of E. Suppose A = 2, B = 1, C = 3, D = 6, and E = 25. The equation would now be set up as:

$$P = \left[\frac{2 + (1 + 3)^2}{6}\right]\sqrt{25}$$

Following the above sequence of operations, the quantities 1 and 3 are first added and then squared, reducing the equation to:

$$P = \left[\frac{2 + 16}{6}\right]\sqrt{25}$$

Further adding 2 to 16 and dividing by 6 results in the quantity 3 inside the brackets. The square root of 25 is 5, which, when multiplied by 3 gives us our final answer of 15:

$$P = [3]5 = 15$$

Subscripts on characters refer to specific applications of that general quantity. V_s, is used to mean velocity at stall or stall velocity. W_s is used to indicate standard weight and ρ_0 is used to mean the density at standard sea level. If the subscript involves several words or terms it may be enclosed in parentheses. Parentheses, in this case, do not mean multiplication.

The Greek letter delta, written Δ, is used to indicate an increment or difference in two values of a quantity. For example, Δh means an increment of altitude. We may speak of a Δh of 100 feet, or we may say that the Δh between 7,200 and 7,300 feet is 100 feet.

Occasionally, we use more than one letter to indicate one quantity, such as TAS for true airspeed or GS for ground speed. Most such terms are abbreviations of the words and are, usually, fairly common terms to pilots. Don't confuse them with terms that are to be multiplied. TAS does ***not*** mean T times A times S, which it would if each of these letters stood for a separate quantity. In Appendix A are standard symbols and subscripts used throughout the book. If in doubt about the meaning of a symbol this table should be consulted.

Presentation of Data

The whole purpose of running performance flight tests is to obtain information on how the airplane performs under various conditions. In order for this information to be of any use, it must be presented in usable form. One of the best ways to present technical data is in the form of a ***line graph***. Line graphs are usually constructed from a series of measurements at discrete points which are plotted and then connected by a smooth line—hence, the name line graph. The line may be straight or have some curvature to it, depending on the relationship.

Figure 2-11 shows a graph of runway distance versus time as measured in a takeoff acceleration test. The points represent actual measurements of the distance from the starting point for various amounts of time measured. At 8 seconds, for example, the airplane was 200 feet from its starting point. The actual points that were measured and recorded are circled. A smooth curve is fitted to these points, however, so that the distance corresponding to any amount of time can be read from this graph. For example, at 11 seconds the airplane is about 370 feet down the runway. Notice that no actual measured point exists, here, however. The graph provides a way of interpolating between actual measurements without additional calculation.

In order to construct the graph in Fig. 2-11, the data had to be measured first. Then, the corresponding points have to be set up in tabular form. Such a table would appear as follows:

Distance (ft)	Time (sec)
100	5.5
200	8.0
300	10.0
400	11.4
500	12.9
600	14.0

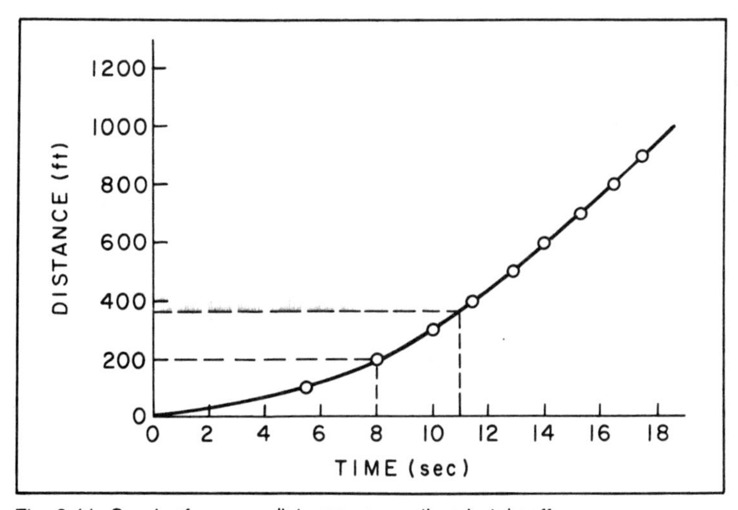

Fig. 2-11. Graph of runway distance versus time in takeoff run.

Distance (ft)	Time (sec)
700	13.3
800	16.5
900	17.6

Next, a suitable scale has to be chosen so that the data will fit the size of the chart. Data must not exceed the boundaries of the chart, yet the scale must be large enough so that the information is readable from the graph. Always mark what each scale represents both in type of information and the units in which it is presented. Here the vertical scale is distance measured in feet and the horizontal scale is time measured in seconds. Usually a graph is referred to by mentioning the vertical scale first, such as a distance-time graph or a distance *versus* time graph.

In the above example, all of the points fell pretty much on a smooth line. This may not always be the case with experimentally measured data. Sometimes unavoidable error in measurement or disturbances in the atmosphere will yield points that completely defy anyone to fit a smooth curve to them. Usually, such a situation indicates a need for more measurements. Figure 2-12 shows rate-of-climb data measured on a slightly turbulent day. A smooth curve

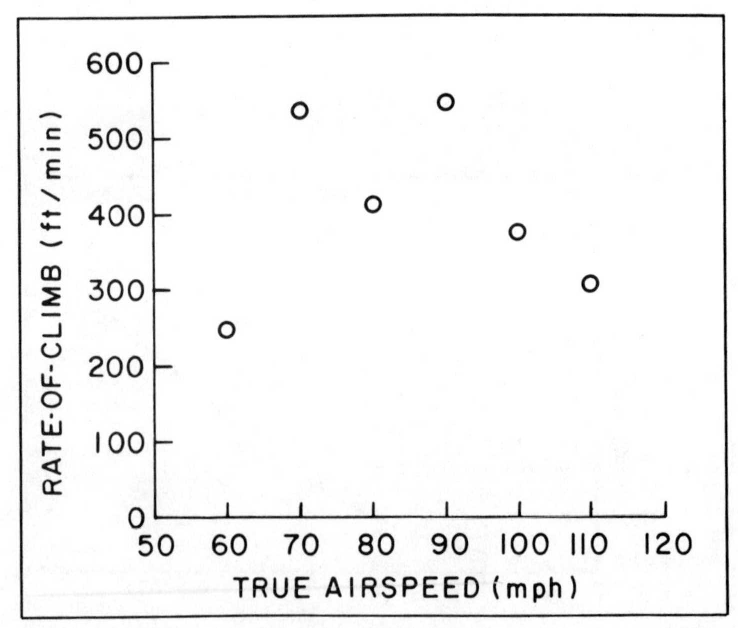

Fig. 2-12. Graph of rate-of-climb versus airspeed: Points measured on turbulent day.

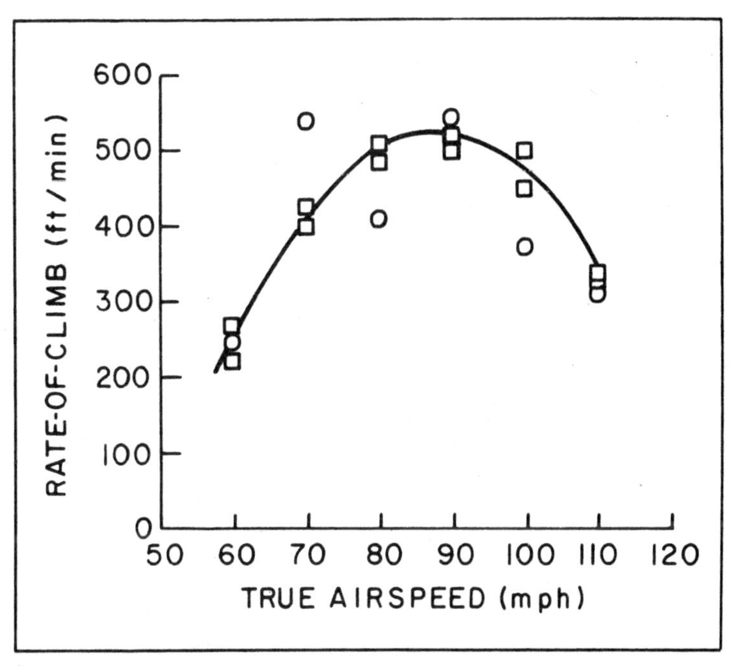

Fig. 2-13. Additional points on a calmer day result in a smooth curve.

is difficult to construct through these points. Additional tests are conducted on a smoother day and plotted in the squares on Fig. 2-13. Since these points are considered more reliable, they are used to construct the curve along with some of the original points (noted in the circles). Notice that some of the original points, those at 70, 80, and 100 mph, are considered so extreme that they are ignored completely in constructing the curve. All of the other points for a given airspeed are pretty much averaged to estimate the true value of rate-of-climb at that speed. It is essential that you have some idea of the general shape of the curve expected. Typical curves for various relationships are shown throughout the descriptions of the tests.

The above discussion involved the plotting of measured data against other measured data. Under such conditions exact measurements are not possible and significant errors will make the plotting of smooth curves difficult. Another way of generating curves, however, is from a mathematical relationship, if one exists. Plotting curves from equations requires much less imagination. Once a standard value is known, often values under other conditions can be calculated. For example, in Chapter 4, an equation is given to determine stall speed for any weight, once it is established for one weight. This equation is:

$$V_{s_2} = V_{s_1} \frac{W_2}{W_1}$$

Suppose that we have measured the stall speed at 2600 lbs and found it to be 65 mph. Stall speed for any other gross weight can then be calculated and plotted as in Fig. 2-14. Calculate as many points as you need to plot a smooth curve. They should all fall right on the line.

Some people find graphs a little confusing and prefer to present data in tabular form. Actually, a table first had to be devised in order to plot Fig. 2-14. The same data as shown on this graph could be presented as in Table 2-1.

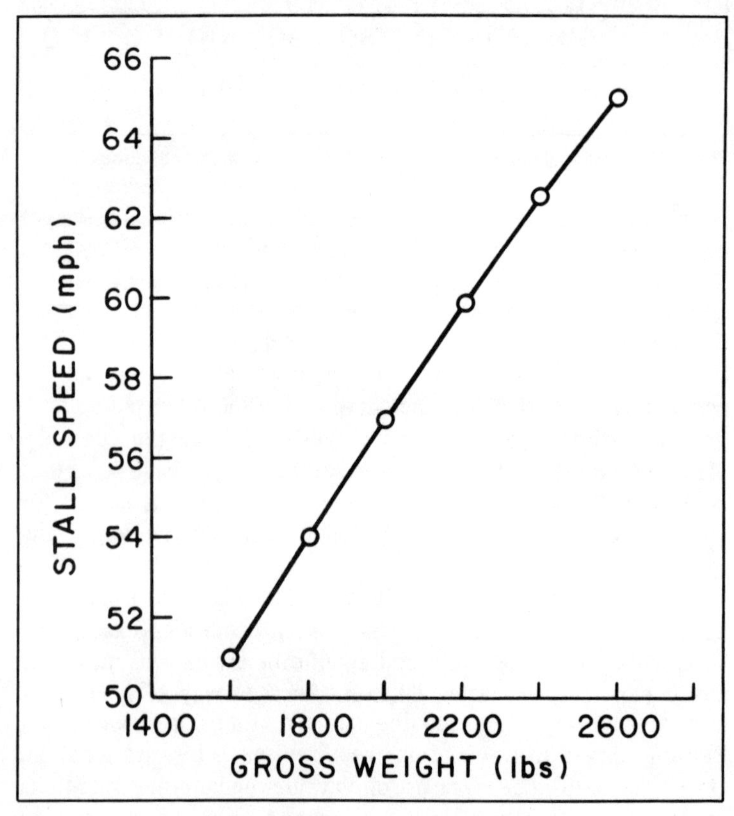

Fig. 2-14. Graph of stalling speed versus gross weight.

Table 2-1. Stall Speed at Various Gross Weights.

Weight (lbs)		Stall Speed (mph)	
2600	2000	65.0	57.0
2400	1800	62.5	54.0
2200	1600	60.0	51.0

Tables are easy to read and yield accurate values as long as you want information corresponding to an exact quantity listed. However, if you require values between these figures, some interpolation is required which could be rather complicated. For example, from the table one could easily note the stall speed for 1800 lbs as 54 mph. However, some interpolation is necessary to determine that it would be about 55.5 mph at 1900 lbs since 1900 lbs is not listed. It would be even more difficult to obtain a value for 1950 lbs. The graph of this same information would yield these values directly, although with some loss in accuracy. The accuracy of information given in graph form depends on the scale at which it is plotted. It does, however, provide for an automatic interpolation process.

Appendix B contains typical performance data as presented in flight handbooks. Graphs of data are given for the Piper Warrior and tables are shown for the Cessna 150. It would be useful to study these to see how such information is presented and to determine which form of presentation is to your liking.

Chapter 3

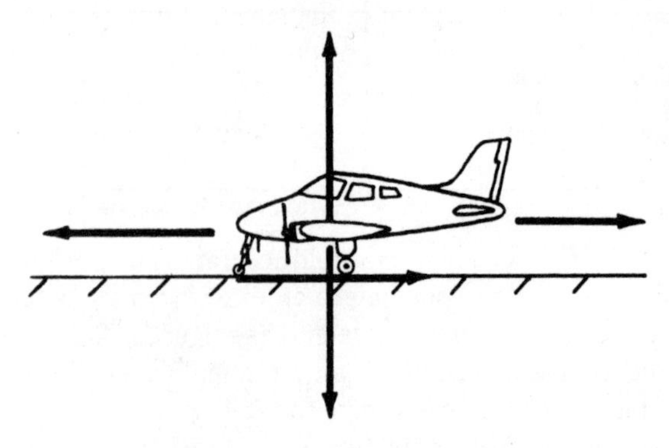

Airspeed Calibration

Practically all airplane performance depends on the airspeed at which the airplane is operating. It is, therefore, necessary to know the true airspeed at which each flight test is performed and to know it rather accurately. The airspeed system may have some deficiencies which cause the indicated airspeed to contain some error. Even though this error may be slight, it is important to determine it and correct the indicated values to what they would be if the airspeed system were perfect. This is known as calibrating the airspeed system and involves establishing a chart (or table) to correct indicated airspeed (IAS) to calibrated airspeed (CAS).

Airspeed Errors

Before going into the actual procedure, let's first discuss the reasons for airspeed indicator inaccuracy. There are two basic types of error. The first is *gauge error*. This is the error inherent in the instrument itself and will be the same regardless of where the instrument is installed. All instruments of every type have inherent error to some degree. Modern airspeed indicators, as is the case with most aircraft instruments, have very little gauge error. This is fortunate because this error requires special instrumentation for accurate measurement.

Gauge error can be determined by use of a device which puts pressure into the pitot tube and also into a very accurate pressure indicator. The two values can then be compared. The airspeed indicator is really a pressure gauge itself, but is marked in values of airspeed rather than pressure readings. Such procedure is shown in Fig. 3-1. This method of checking gauge error should be done only by someone familiar with both the equipment and the procedure. It is quite easy to damage the airspeed indicator by putting too sudden or too much pressure into it. You probably recall your flight instructor telling you never to blow into the pitot tube. This is why. If you suspect a significant gauge error, it is wise to have it checked by a qualified instrument repair shop.

The second type of error is called *position error*. This error is due to the location of the pitot and static ports in the pressure field surrounding the airplane. This error will be different for the same indicator installed in different airplanes. The pitot tube is aimed directly into the airstream and picks up the total pressure; that is, the dynamic pressure, or that pressure from the moving air plus the pressure of the surrounding still air. The static port is meant to measure only the static or still air pressure. The airspeed indicator operates by differentiating between these two pressures and reading out only the dynamic pressure. Obviously, if the pressure at the static port is not quite the true static pressure, then the difference the indicator measures will be wrong and an airspeed error exists.

Pitot tube and static port locations are designed with considerable care to minimize error. However, the error usually will not be consistent, since any change in attitude of the airplane causes a change in the pressure field surrounding it. The pitot tube is usually located under the wing because here the flow is nearly always

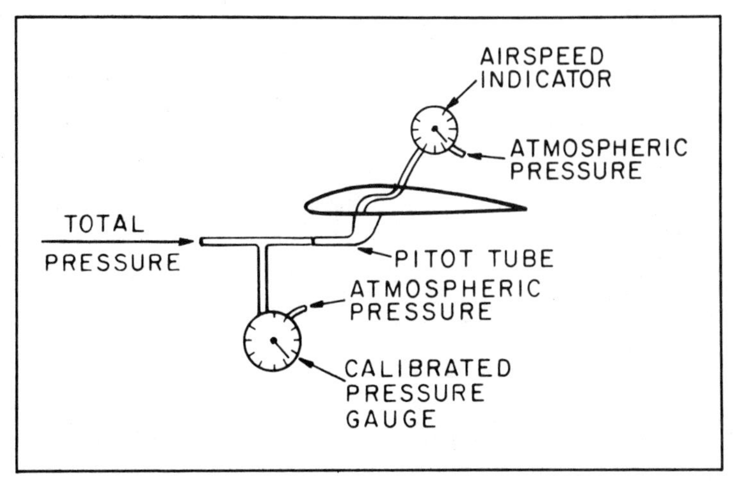

Fig. 3-1. Gauge error test set-up.

Fig. 3-2. Typical pitot tube location beneath wing.

aligned with the wing surface regardless of the angle of attack. Figure 3-2 shows such an installation. This method of mounting usually all but eliminates error in the pitot tube.

The static port is harder to locate. It sometimes is located with the pitot tube or somewhere on the aft portion of the fuselage. It may incorporate some sort of baffle to minimize the effect of change in air direction. However, as the angle of attack changes slightly, it may pick up a little effect from dynamic pressure. This amount can vary with various angles of attack. Figure 3-3 shows a typical static port on the fuselage.

Test Procedure

Because the airspeed is dependent upon angle of attack, position error probably will vary with different airspeeds. Therefore, it

Fig. 3-3. Typical static port location on aft portion of fuselage.

is necessary to check the error for a number of airspeed values throughout the entire range normally flown. Position error alone is difficult to determine accurately without knowing the exact gauge error. The entire system error, however, which includes both gauge and position error, can be determined rather easily.

The first step is to select a calibration range. This means choosing two straight, parallel landmarks a known distance apart. Roads, railroads, fences, power lines or any straight, prominent objects can be used. If you don't know the distance between the objects, measure it. A calibration range should be about one to five miles in length, though it doesn't have to be an even number of miles. Just be sure you know the *exact* distance. Figure 3-4 shows a typical range. The fenceline and the highway parallel to it form two boundaries 3.5 miles apart. Notice that the road running perpendicular to these landmarks crosses both of them and therefore can be used to measure the distance. This is best done with a tape measure but can be approximated by using an automobile with an accurate odometer. Quite often an airport will have a known calibration range nearby, so ask your FBO or some of the old-time pilots around the field. This will save you a lot of effort, because the

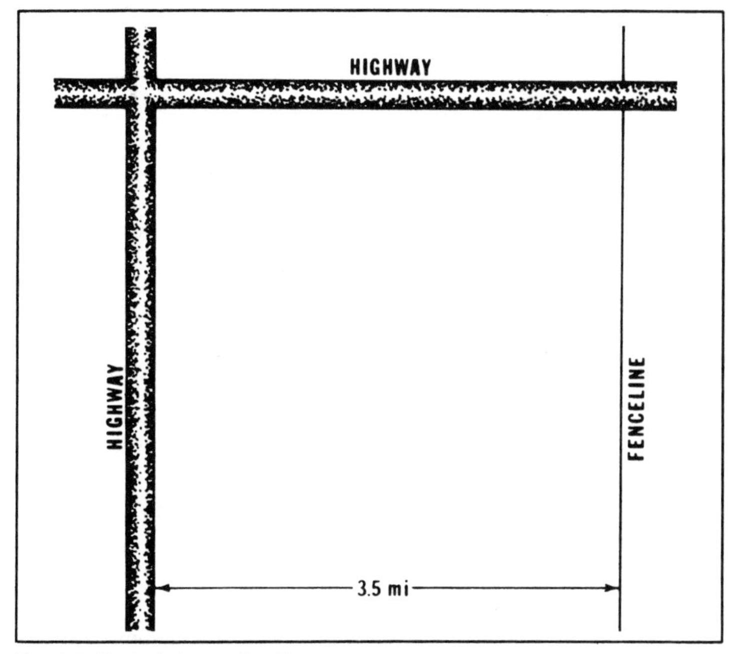

Fig. 3-4. Typical airspeed calibration range.

biggest part of this procedure is selecting and measuring a test range.

The only other equipment you need is a stopwatch. The basic procedure simply is to fly across the range at a constant airspeed and record the time it takes to go this known distance. This would be the complete test if there were no wind. The true airspeed would equal the groundspeed which would be the distance divided by the time. It could be compared with the airspeed value read from the indicator (corrected to TAS) and the error determined. However, as all pilots well know, wind nearly always is present and usually is the hardest factor to determine accurately in time-speed-distance problems. Well, don't give up yet, because the wind problem can be solved very easily for our purposes here.

Suppose we had a direct headwind over our course. The wind would subtract from our groundspeed by the amount of the wind speed. Now, suppose we turned around and flew the course in the opposite direction. The wind would now add to our groundspeed by this same amount. Thus, if we flew the course in both directions, and averaged our ground-speed, we would get the true airspeed. For example, suppose the airplane had a true airspeed of 100 mph and the wind were 20 mph. Flying from landmark A to landmark B the groundspeed would be the TAS minus the windspeed or 80 mph. Flying in the opposite direction, the wind would create a tailwind and the groundspeed would be 100 mph plus the windspeed or 120 mph. Now if we average these two groundspeeds, that is add them and divide by two, we end up with the original true airspeed of 100 mph. Therefore, if we can measure the groundspeed in both directions, we can calculate the true airspeed. Groundspeed is easily determined as the distance divided by the time to go that distance. This method is shown in Fig. 3-5. The procedure is:

1. Calculate groundspeed from A to B.

$$\text{GS (A} - \text{B)} = \frac{\text{distance in miles}}{\text{time in minutes}} (60)$$

2. Calculate groundspeed from B to A.

$$\text{GS (B} - \text{A)} = \frac{\text{distance in miles}}{\text{time in minutes}} (60)$$

3. Calculate the average groundspeed by adding the two groundspeeds and dividing by 2.

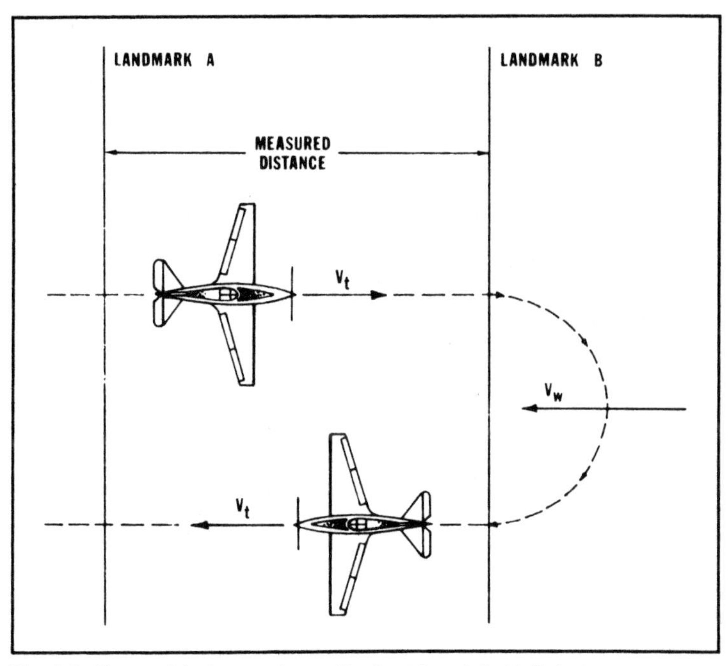

Fig. 3-5. Airspeed test procedure with direct headwind-tailwind.

$$\frac{GS\ (A - B) + GS\ (B - A)}{2}$$

Steps 1 and 2 can also be performed by use of a flight computer.

Now, let us consider the case where we have a wind with a crosswind component as in Fig. 3-6. If we keep the airplane headed in the direction A-B, just as we did for the no-wind or direct headwind-tailwind case, and neglect the drift, we will keep the true velocity vector headed in the proper direction. The only part of the wind that will affect the groundspeed will be the headwind-tailwind component as indicated on the figure. The drift component will not affect our groundspeed since it is perpendicular to our heading. It will have the effect of causing the airplane to drift so that it does not actually arrive at point B but rather at point C. This position is of no consequence to the problem, however. The procedure, therefore, is exactly the same as before.

If you don't follow all of the mathematics, the procedure is very simple. Just head the airplane perpendicular to your landmark lines, neglect the drift, and fly a constant heading, constant airspeed course, recording the time it takes to fly from one line to the other. Then fly a reciprocal heading at the same indicated airspeed and again record the time. Calculate each groundspeed by dividing the distance by the time, add the two groundspeeds, and divide by two. The result is your true airspeed. Now, change your measured true airspeed value to calibrated airspeed. This can be done with your flight computer in just the same way you make TAS corrections, except that you read a calibrated value from the inner scale for a given value of true airspeed on the outer scale. This calibrated value can then be compared with your indicated value read from the gauge to determine your error. If you do this for a number of different indicated airspeeds, you could then plot a chart of indicated versus calibrated airspeed. Figure 3-7 is such a chart, and it shows, for example, that for an indicated value of 85 mph your calibrated or actual value is only 80 mph. This chart resulted from actual flight tests by the author on a modified Globe Swift.

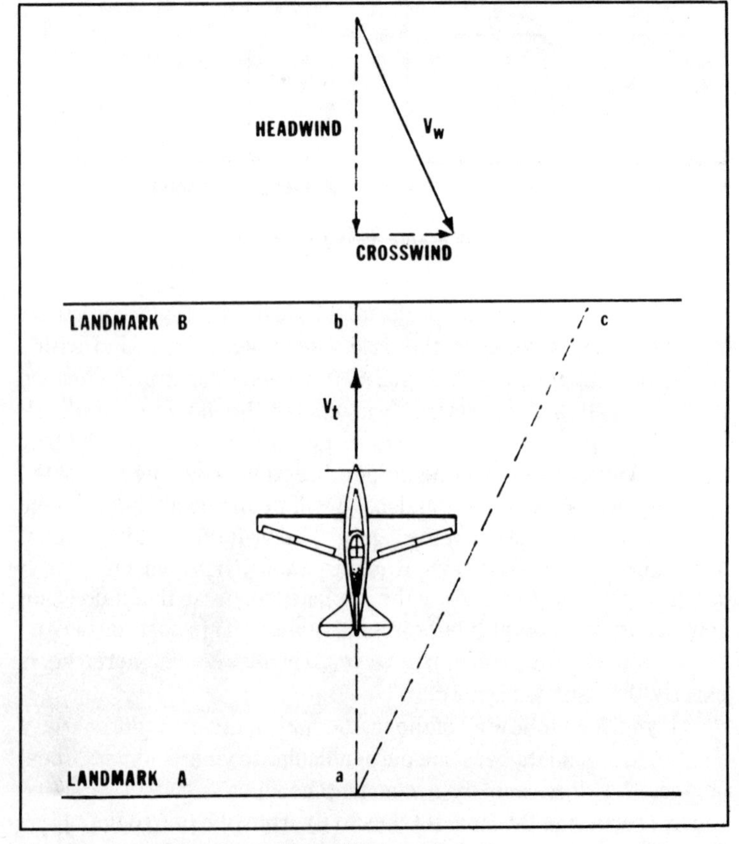

Fig. 3-6. Airspeed test procedure with crosswind.

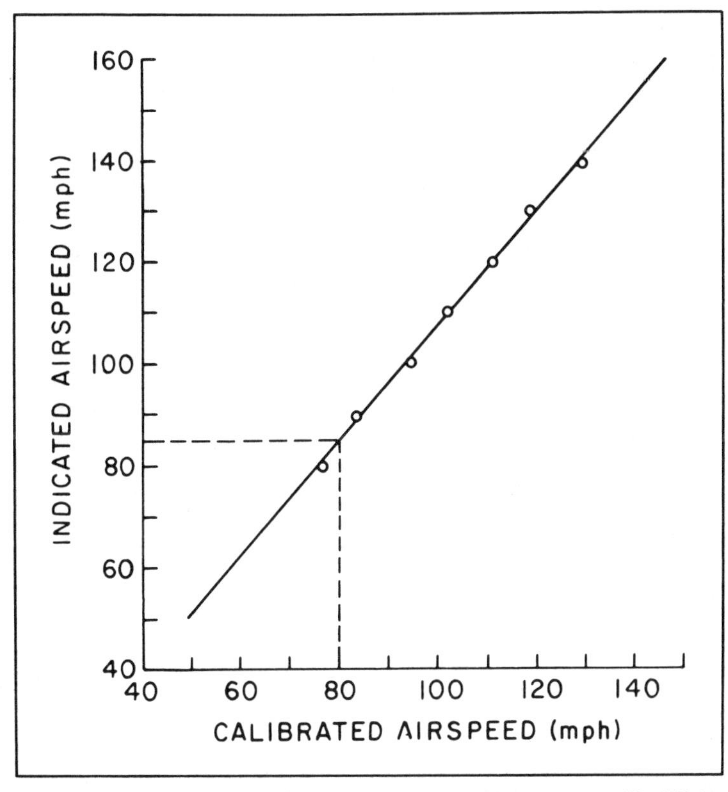

Fig. 3-7. Airspeed calibration chart as obtained from flight tests of modified Globe Swift.

Fig. 3-8. Using wing leading edge as a guide to crossing airspeed range landmarks with low-wing.

It usually is best to utilize two people in conducting the tests. In this way the pilot can concentrate on holding a constant heading and airspeed while the observer can sight the landmarks and operate the stopwatch. The observer should sight over some fixed point on the aircraft, such as a point on the leading edge for a low-wing (Fig. 3-8) or a point on the strut or landing gear for a high-wing (Fig. 3-9). It is advisable to start the run some distance from the first mark so the airplane can be trimmed out, and its power set up for a constant speed and altitude, before the timing is started. A constant altitude is desirable during the run but slight changes such as a hundred feet or so will not alter your measurements appreciably. Make your runs as low as regulations or safety precautions permit for the best accuracy in timing. Of course, a very low amount of atmospheric turbulence is a must for any degree of accuracy.

True Airspeed

As mentioned previously, the airspeed indicator is really a pressure gauge which measures the dynamic pressure of the air. This pressure depends on the air density and is equal to $\frac{1}{2} \rho V^2$, that is, ½ times the density times the velocity squared. Since the indicator *reads* in values of velocity, however, some constant value of density must be used. That value is standard sea level density. This means that the indicator, even in the absence of any error, never reads true airspeed except when the density is at standard sea level. Thus, to get true airspeed (TAS) we must make a correction for density anytime it is different from that at sea level. Since this is most of the time, calibrated airspeed normally has to be corrected to determine TAS.

Fig. 3-9. Using wheelpant with high-wing as guide to crossing landmark.

It was already mentioned before that corrections between CAS and TAS can be made by use of a flight computer. To do this you must set in the proper pressure altitude and temperature. These are the quantities needed to measure density and what the computer is really doing is calculating the density and its difference from standard sea level density. You can also make this correction without a computer by use of the following equation:

$$\text{TAS} = \frac{\text{CAS}}{\sqrt{\rho/\rho_0}}$$

ρ/ρ_0 is the density ratio and can be determined from Figs. 1-3 and 1-4 by using the proper temperature and pressure altitude. Dividing the square root of this ratio into the calibrated airspeed yields the true airspeed.

The normal procedure to determine true airspeed is then done in two steps:

1. First, read a value of IAS from the airspeed indicator and correct it to CAS by the calibration curve established in the test in this chapter.

2. Correct the value of CAS to TAS by applying the temperature and pressure altitude (at the time of the test) to either a flight computer, or to the charts to obtain ρ/ρ_0 and then use the above equation.

True airspeed is normally required when determining performance involving the time to cover a certain distance. The time to ETA, for example involves a determination of groundspeed, which, in turn, depends on TAS. Thus, cruise performance is usually established for true airspeeds. Takeoff performance also relies on TAS, since the time to cover a certain distance on the ground is used to measure the acceleration of the airplane.

On the other hand, certain qualities of the airplane depend on dynamic pressure. Since this is what the airspeed indicator measures, indicated airspeed is really a better determination of such performance. Stall speed, for example, depends on lift, which is dependent on dynamic pressure. The IAS at which the airplane stalls is, thus, the important figure. Climb performance, also, is usually determined for IAS, since this is what the pilot is reading in order to establish a climb speed.

Airspeed Calibration Example

Conditions:

☐ Distance between landmarks (A to B) = 3.0 miles.

☐ Time to fly from A to B = 2.3 minutes.
☐ Time to fly from B to A = 2.1 minutes.
☐ Temperature = +20°C (68°F).
☐ Pressure altitude = 1000 ft.
☐ Indicated airspeed = 85 mph.

1. First, calculate groundspeed from A to B.

$$GS = \frac{3 \text{ mi}}{2.3 \text{ min}} \times 60 = 78.26 \text{ mph}$$

2. Second, calculate groundspeed in the opposite direction (B to A).

$$GS = \frac{3 \text{ mi}}{2.1 \text{ min}} \times 60 = 85.71 \text{ mph}$$

3. Now average the groundspeed.

$$GS \text{ (average)} = \frac{78.26 + 85.71}{2} = 81.99 \text{ mph}$$

Round this off to 82 mph.

4. This is your true airspeed. Convert this to calibrated airspeed (CAS). Set 20°C opposite 1000 ft. pressure altitude on your flight computer as shown in Fig. 3-10. Opposite 82 mph TAS on the outer scale read 80 mph CAS on the inner scale (Fig. 3-11).

Fig. 3-10. Setting 20°C temperature against 1000 ft. pressure altitude on flight computer.

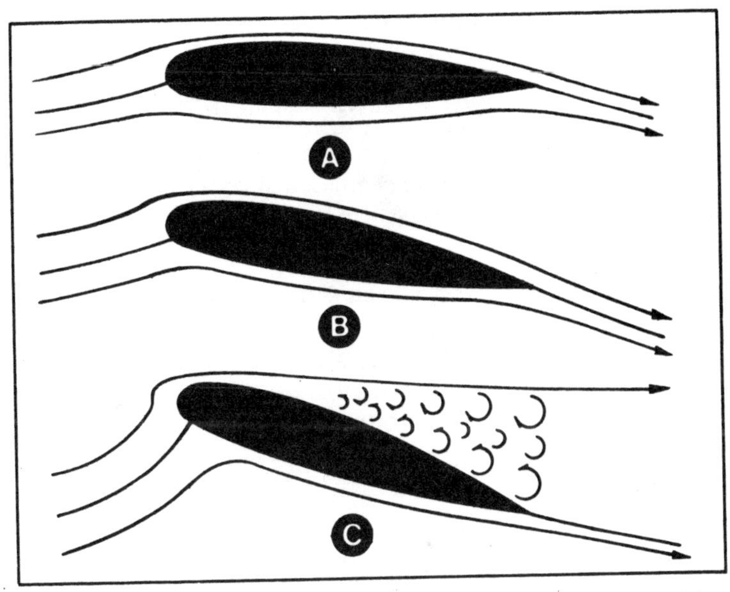

Fig. 4-1. Stall sequence of airfoil. (A) At normal angle-of-attack with stagnation point near leading edge. (B) At higher angle of attack with stagnation point moved back on lower surface. (C) At stall angle showing separated flow over top surface.

Eventually, however, at a very high angle of attack, this path gets so long that the air doesn't have enough energy to travel this entire path. Remember that it is being retarded by friction along the surface and the longer the path, the greater will be this frictional effect. At the limit, it breaks away from the top surface. The result is a flow pattern as shown in Fig. 4-1C. The flow over the top surface is separated from the leading edge back. With no smooth flow over the top surface there is nothing to lower the pressure and the lift drops off significantly. This condition is what we call a "stall."

Lift does not depend upon angle of attack alone, but is also affected by the density of the air, the velocity, and the wing area. Mathematically the equation is:

$$L = K \alpha \rho V^2 S$$

where

L = lift
K = constant of proportionality
α = angle of attack
ρ = density
V = velocity
S = wing area

This equation says that it is the product of all of these factors multiplied together, including the square of velocity, that determines lift. The constant, K, is a number depending on the units of the various factors and also the wing and airfoil shape. At a given altitude (constant density) and with a given wing area, the two remaining variables are angle of attack and velocity. Including the density and wing area in the constant would then make the equation:

$$L = K \alpha V^2$$

Thus, to maintain a certain amount of lift we can trade off velocity for angle of attack. This is why at low airspeeds we need to have a high angle of attack, and at high speed, a low angle of attack. Furthermore, from a balance of the four basic forces on the airplane (lift, weight, thrust, and drag), the lift must equal the weight in steady level flight. Therefore, we can say:

$$L = W = K \alpha V^2$$

where W is the gross weight of the airplane.

If we plot lift versus angle of attack for a certain airfoil section, it looks like the chart in Fig. 4-2. Notice that the lift goes up proportionally to angle of attack until the stall angle is reached. Again, for a given airfoil there is a *fixed* angle of attack at which it

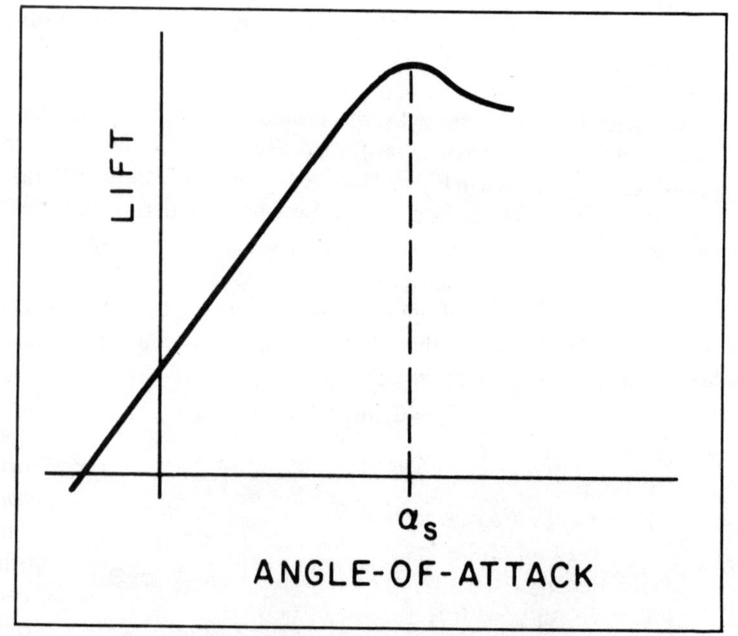

Fig. 4-2. Lift versus angle-of-attack showing location of stall angle.

stalls. Therefore, at the stall point, we can include the angle of attack in our constant and the lift equation is:

$$L = W = K\,V^2_s$$

This equation shows that the higher the weight, the higher our stall velocity must be. For this reason we need to consider the stall speed at various gross weights.

Another factor that affects stall speed is the flap setting. In our original equation for lift, a factor (K) was included to account for, among other things, the airfoil shape. Lowering flaps changes the airfoil shape and thus alters K. It turns out that for higher flap deflection angle, this constant becomes higher and thus, more lift is obtained. All of this means that another set of stall speeds will occur for various weights with a certain degree of flap deflection. Hence, to get a complete set of stall data, stall speed must be measured with various angles of flap deflection.

Fortunately, different altitudes do not affect stall speed if we consider only calibrated airspeed (CAS). The *true* airspeed at which an airplane stalls is different at different altitudes for the same weight and flap setting. However, the density change with altitude affects the airspeed indicator proportionally to the way it affects stall speed. If stall speed were measured at sea level, for example, and then several thousand feet above sea level, the lower density at the higher altitude would cause the airplane to stall at a higher TAS. However, the lower density would also reduce the dynamic pressure in the pitot tube and the indicator would read no higher that at sea level (assuming the instrument error to be constant). This is the reason you normally have to calculate TAS from an indicated value.

Bank Effects on Stall

When the airplane is flying straight and level, the lift being generated by the wing is equal to the weight. The lift force is perpendicular to the plane of the wing and, in this case, this plane is parallel to the horizon. However, when the airplane is banked, the lift is still perpendicular to the plane of the wing, but now the wing is tilted with respect to the horizon. The lift force is, therefore, tilted. Weight is still the same and always tends to pull vertically downward. Thus, more lift must be generated so that the *vertical component* of the lift vector equals the weight. In addition, the airplane in a bank is experiencing a centrifugal force due to the turning of the airplane. Part of the lift must be utilized to overcome this centrifugal force. The horizontal component of the lift provides this force.

For these reasons you can see that additional lift is required to maintain level flight in a bank. Stall speed is proportional to the square root of lift, so that greater lift results in higher stall speed. Remember that in level flight, higher gross weight can result in higher stall speed for the same reason. Higher weight requires higher lift to overcome it. For whatever reason, when more lift than normal exists, stall speed will be higher. Therefore when banking is done, the stall speed goes up. Steeper bank requires more lift; consequently, higher stall speed will result. The amount of increase with bank angle is the same for all airplanes. Therefore, the stall speed for any bank angle can easily be calculated by use of Fig. 4-3. An airplane that stalls at 50 knots in level flight, for example, would stall at 1.25 times that speed in a 50° bank, or at 62.5 knots.

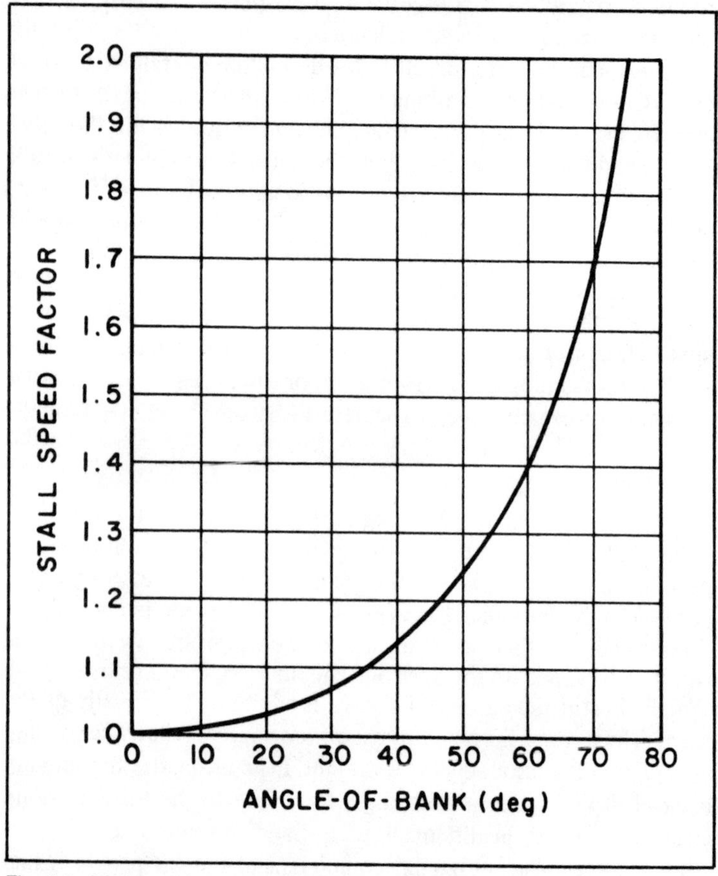

Fig. 4-3. Stall speed factor versus angle of bank.

Fig. 4-4. Stalling the airplane very gradually with nose nearly level.

Test Procedure

The actual flight testing of stall speed is relatively simple. Since the test does depend on weight, it is necessary to determine the gross weight rather accurately. The airplane should then be flown to a safe altitude and area in which to perform stalls. Since calibrated airspeed for stall does not vary with altitude, the exact altitude at which the tests are conducted does not matter.

The stall should be performed power off and entered very gradually. Abrupt entry will result in an accelerated condition and not yield accurate results. It is best to keep the airplane in fairly level attitude (nose on or slightly above the horizon) and definitely keep the wings level (Fig. 4-4). Very slowly pull back on the stick and watch the airspeed indicator. The needle should slowly and smoothly move down and stop just at the break of the stall. Record this velocity. The procedure should then be repeated at various flap settings. If definite notches of flap settings are incorporated in a manual controlled system, such as in many Pipers, measure the stall speed for each setting. If a continuous motor-driven system is employed, as in most Cessnas, measure full deflection and as many other settings as you desire. For a retractable gear airplane, you should also measure stall speed in both gear-up and gear-down configurations. The FAA specifies a stall speed for landing configuration known as V_{s_0}. This is the stall speed with engines idling, gear extended, propellers in takeoff position, and flaps in landing configuration. FAR Part 23 requires all single-engine airplanes and certain light twins to have a stall speed in this configuration of no greater than 61 knots (70 mph).

To determine the stall speed at other weights, the gross weight could be changed by changing the loading and the experiment repeated. However, once the stall speed is known for one weight, it can be calculated for other weights by the following equation:

$$V_{S_2} = V_{S_1}\sqrt{\frac{W_2}{W_1}}$$

This equation says that if you know the stall speed, (V_{S_1}), for one gross weight (W_1), then the stall speed for another weight (V_{S_2}) can be determined by multiplying V_{S_1} by the square root of the new gross weight (W_2) divided by the original gross weight. This same situation holds true for the various flap settings, also.

In any case, the values you obtain are in indicated airspeed (IAS). They will have most meaning to the pilot if plotted that way, that is, IAS versus gross weight for various flap settings. Figure 4-5 shows such a chart.

If you need the true airspeed value of stall speed, the IAS value must first be corrected to CAS by the chart constructed in Chapter 3 and then converted to TAS by correcting for appropriate altitude and temperature. This procedure can be done with the flight computer as discussed also in Chapter 2.

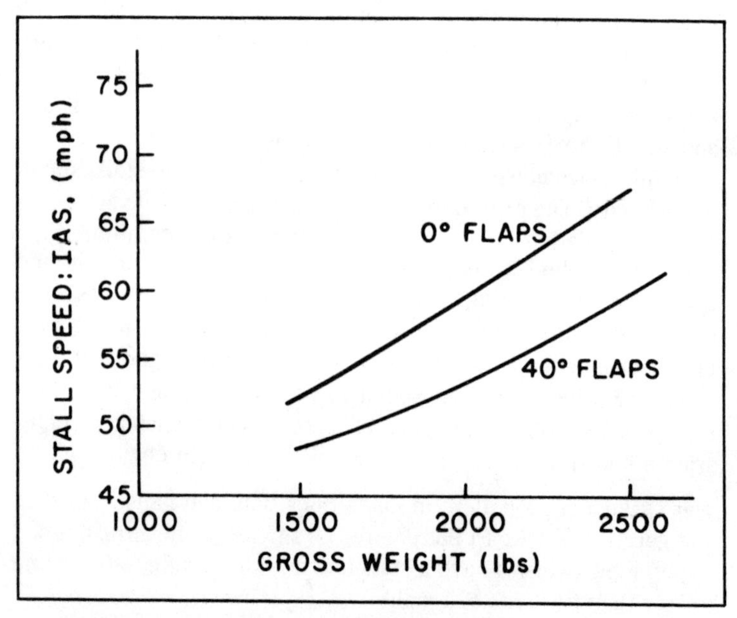

Fig. 4-5. Typical chart of stall speed versus gross weight for various degrees of flap deflection.

Stall speed for any angle of bank can also be plotted into a chart by taking the appropriate value of stall speed for wings level and multiplying by the various factor for the respective bank angles from Fig. 4-3.

Stall Speed Test Example

Test Conditions:

- ☐ Gross Weight = 2430 lbs
- ☐ Flaps = 0°
- ☐ Stall speed = 67 mph IAS

1. Determine stall speed at 2500 lbs, use the weight correction equation.

$$V_{S_{2500}} = V_{S_{2430}} \sqrt{\frac{2500}{2430}}$$

$$= 67 \sqrt{\frac{2500}{2430}}$$

$$= 68 \text{ mph}$$

2. To determine stall speed at 2000 lbs, use the same equation.

$$V_{S_{2000}} = 67 \sqrt{\frac{2000}{2430}}$$

$$= 61 \text{ mph}$$

Note that the airspeeds are rounded off to the nearest whole mile per hour in these examples.

Chapter 5

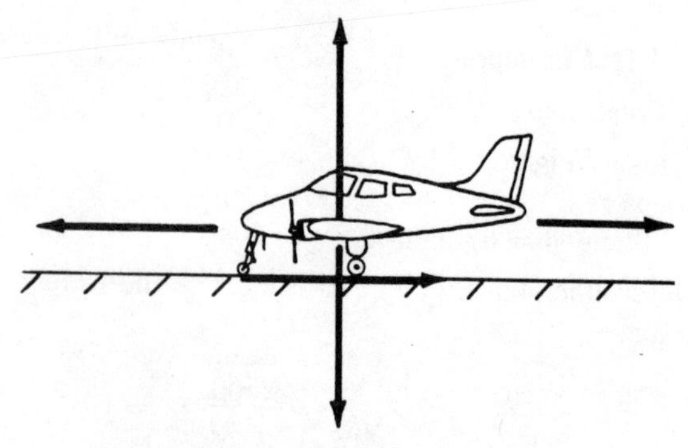

Takeoff Performance

The whole purpose of a takeoff run is to get the airplane up to flying speed. The airplane will fly at any speed above stall. However, to lift off at just one or two knots above stall is a bit risky. Just the slightest increase in angle of attack would cause the airplane to stall and drop back onto the runway. In addition to being an unsafe operation, this practice would result in a run longer than necessary to ultimately get airborne. Therefore, some additional margin of speed is added to the stall speed to arrive at what is considered both a practical and a safe takeoff speed. Usually, 20 percent above stall is considered an adequate margin of speed. Thus, if your stall speed is 60 knots, 72 knots would be considered takeoff speed. Remember that this stall speed is for the configuration (gear and flap setting) and weight at which you are operating.

The problem in takeoff performance is to determine the runway distance required. The actual distance for a takeoff is highly dependent on pilot technique. One pilot may use up a lot more runway for takeoff than another under the same conditions and with the same airplane. An equitable measure of takeoff distance, then, is that distance required to reach a specified value of takeoff speed. Whether one actually lifts off right at that distance and speed is immaterial. The important fact is that at this point it *could* be lifted off.

Takeoff performance is now reduced to the problem of determining the distance required to accelerate from zero velocity (at the beginning of the run) up to takeoff velocity. If we use the value of 1.2

V_s for takeoff velocity, the situation is further defined in terms of specific numbers.

Forces Acting

How fast the airplane accelerates depends on the quantity of the forces acting. Normally, in flight we have drag tending to retard our forward progress and thrust to overcome this force. We also have these forces acting on the airplane during takeoff, but in addition we have the frictional force of the tires rolling on the runway. This force is alleviated, somewhat, as the takeoff run progresses by the lift reducing the weight on the wheels. This effect is rather slight, however, for many modern tricycle-gear airplanes, unless a definite effort is made to raise the nose slightly (and, hence, increase angle-of-attack) early in the run. Too much angle-of-attack, on the other hand, increases drag and defeats the purpose of reduced friction.

Another force acting to retard forward motion is the internal force of the airplane. This force is explained by Newton's First Law of Motion, which says that "bodies at rest tend to remain at rest." All of these forces are shown in Fig. 5-1. Notice that drag, friction, and inertia all tend to retard the forward motion and, hence, acceleration. The only force acting for us is the thrust. Drag, friction, and inertial forces can all be lumped together and called "retarding force." A greater difference between thrust and retarding force will result in faster acceleration. Both frictional force and inertia are proportional to weight, so lower weight means lower retarding force, and hence, faster acceleration. Of course, faster acceleration means the ability to reach takeoff speed in a shorter distance.

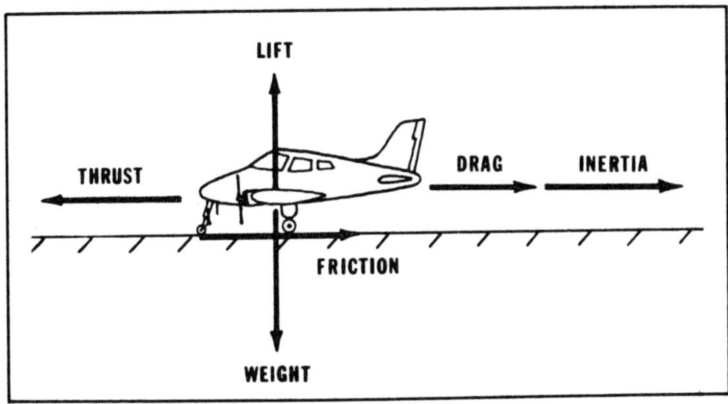

Fig. 5-1. Forces acting on the airplane during takeoff run.

Airports are ideally laid out on a perfectly flat surface. Since this is not always possible, runways usually have some amount of slope to them. If a fair amount of slope exists in a runway, then an additional gravity force is present during takeoff. An upslope run means a gravity force tending to retard motion and a downslope run results in a gravity force which adds to the acceleration. Since standard takeoff data is desired for a no-slope condition, it is necessary to perform tests at an airport with a relatively level runway. Runway gradients (slopes) of 0.3 percent or greater are supposed to be indicated in the Airport/Facility Directory. However, not all airport listings include this information. If available, it is best to consult the engineering drawings for the airport construction. The airport manager should have such drawings. If you are operating from a very small field, such information may not exist. If you are uncertain of the slope of your strip or, if it has appreciable slope, it is wise to go to another airport where conditions are known to be better to run the takeoff tests.

Surface conditions of the runway determine the amount of retarding frictional force. Grass runways have more friction than hard-surfaced runways, for example. Thus, the data obtained will be for the type of surface that you use in the tests. If you want takeoff distance for a grass strip, you will have to run the tests on a grass strip. If you are based at such a strip and want takeoff performance for a hard-surfaced runway, then, again, you will have to find such a facility at which to perform the tests.

Test Procedure

Since the takeoff problem has been explained as one involving the acceleration of the airplane, a *possible* takeoff distance can be determined by measuring how long a distance it takes to accelerate to takeoff velocity. Such a test of distance to accelerate to a certain speed could also be applied to an automobile or any other land vehicle. Thus, it should be obvious that the test procedure does not involve a noting of the point of liftoff. In fact, it is not necessary to even take off at all in order to determine where the airplane *could* take off. All that is necessary is a takeoff run keeping track of the time to cover various increments of distance during the run. Usually, the run will be of necessary length so that not enough stopping distance is left on the runway. For this reason, it is wise to go ahead and complete an actual takeoff. This procedure also assures that you have continued the run long enough to reach takeoff speed. The point is that with a sufficiently long runway, an actual takeoff does not have to be made to perform the test.

Prior to starting the test runs, some markers must be located along the runway at spacing of about 100 to 200 feet. Existing runway lights serve well as distance markers. Their spacing can usually be found on the airport layout drawings. If not, the distance between them can be measured with a tape measure. If your runway has no lights, wooden stakes can be placed every 100 feet for the first 300 or 400 feet and, then, every 200 feet up to about 1500 or 2000 feet. Be sure to get the airport owner's permission before doing this.

Load the airplane to approximately the standard weight value that you have decided upon. There are weight corrections that can be applied for tests at nonstandard weight, but it is best to operate at the weight for which data is desired.

To begin the test run, taxi up to the first marker. The pilot should very slowly ease up to this point until the spotter just has the first marker spotted as shown in Fig. 5-2. Spotting is done by sighting along a point on the airplane to a marker on the ground. The spotter should keep his head in a fixed position so that the angle of sighting does not vary. If the airplane is a low-wing, the leading edge of the wing tip or tip tank is a good spotting point. For a high wing airplane a point on the wing strut, a pitot tube or some other protrusion below the wing can be used. The right front seat is usually the best position for the spotter with the markers laid out on the right side of the runway.

Fig. 5-2. Spotting the runway lights to be used as takeoff distance markers.

With the airplane lined up at the first marker, considered the "zero point," the pilot should then apply full power while holding the brakes. When the brakes are released, start the timing. This procedure does not necessarily shorten the takeoff run. It merely ensures that the run is started at a definite point in time. The airplane should then be allowed to accelerate in a normal manner down the runway. Don't attempt to lift the nose prematurely. If the airplane is a taildragger, raise the tail as soon as possible to the normal takeoff position. It is important to keep the airplane headed as straight as possible down the runway so that the wing passes each marker at the same angle.

As each marker is passed, record the time. There are several ways in which to do this timing. The simplest way is, probably, to use a stopwatch which has a "lap" feature. This is one in which the reading can be stopped momentarily, but the actual timing continues from the initial starting point. This procedure allows the time at one marker to be read and recorded and then the original time switched back on so that the procedure can be repeated at each successive marker.

An alternate procedure is to use a portable tape recorder instead of a stopwatch. Start the recorder prior to release of the airplane and call "start" or "go" into the recorder as the run begins. Then, as each marker is passed, call "one," "two," "three," and so on, into the recorder. Be sure to speak in a short, abrupt manner so that a distinct time of marker passing can be determined. Then, after the flight, the recording can be played back and the time to each marker determined with a stopwatch. Another method is to photograph a stopwatch at the instant each marker is passed by use of a rapid-wind camera. The timing is then read from the photographs after developing. Both of these methods require some additional time and effort after the flight tests.

As a last resort, a single continuously running stopwatch can be employed. For this method one person has to continually monitor the stopwatch and catch the reading as the spotter calls out the marker. This works pretty well for the first few markers, but as the acceleration builds up, the last few are passed pretty rapidly. You might try several methods, or original variations of them, to see which works best for you.

In any case, the time to pass each marker must be obtained. This time is then plotted for the respective distances of the markers as shown in Fig. 5-3. The average speed between each pair of markers is then determined by calculation. Since average speed is,

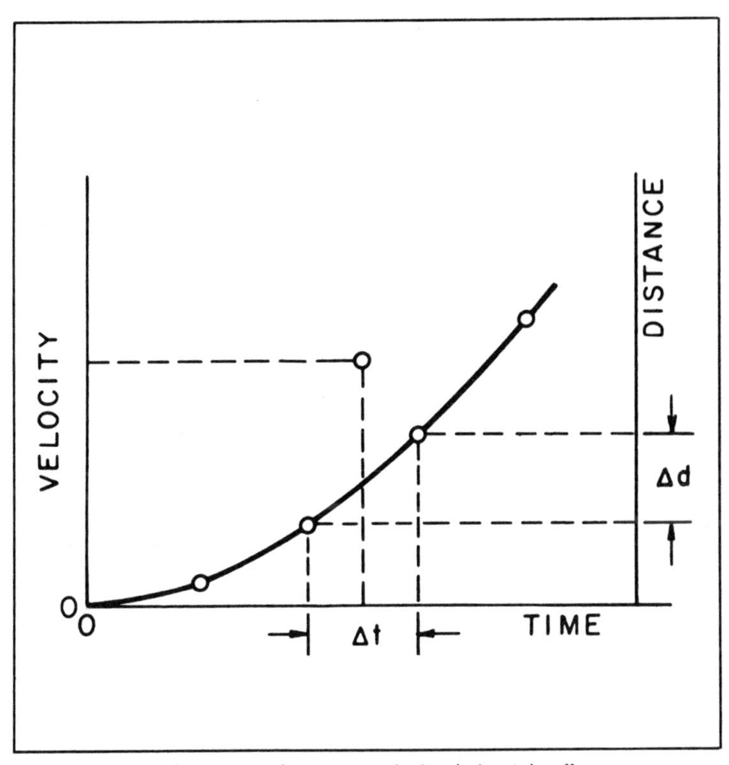

Fig. 5-3. Plotting distance and average velocity during takeoff run.

simply, distance divided by time, take the distance between two markers, as shown, and divide by the difference in time between these same markers. Establish a velocity scale on the left hand side of the chart and plot the average velocity at a point halfway between the marker positions. The equation for velocity is:

$$V_{average} = \frac{\Delta d}{\Delta t}$$

If the distance is measured in feet and the time in seconds, then the velocity will be in feet per second. To convert to mph, this value must be divided by 1.467. Thus:

$$V_{av}\ (\text{mph}) = \frac{\Delta d}{\Delta t\ (1.467)}$$

To convert to knots divide by 1.687 instead of 1.467. After computing the average velocities between each set of markers and plotting them, draw a smooth curve, fitted as closely as possible to both the distance points and the velocity points.

If the velocity points calculated by this method do not fall on a smooth curve, a better procedure is outlined as follows: Plot the points for the time to various distance markers, as in the above procedure, and draw a smooth curve through these points. Then, take even increments of time, for example, every two seconds and find the distances corresponding to these times from the curve drawn. Now, use the difference in these distances as the Δd in the above equation and the Δt that you have selected (such as the two seconds mentioned) to calculate average velocities. The smaller the time increment chosen, the more accurate the velocity curve will be.

Now determine your possible takeoff speed as 1.2 times the stall speed. This value must be TAS, so take the indicated stall speed for the configuration being flown (gear down, we hope), correct it first to CAS by your calibration curve, and then correct this for density to TAS. Note that this procedure requires the stall speed to first be determined by methods in the previous chapter and an airspeed calibration to be done as outlined in Chapter 3.

Once the takeoff speed in TAS is determined, enter the chart you have constructed at this value on the velocity scale (left hand side). Move right until you hit the velocity curve. Then go vertically up or down as necessary to hit the distance scale. Now read the distance scale directly to the right of this point. This is your takeoff distance. The procedure is shown in Fig. 5-4.

Wind Correction

Tests should be run in wind as near to dead calm as possible. Light headwinds of 10 knots or less can be tolerated and corrected for as long as they are steady. Some means of measuring the wind must be employed. There are a number of handheld anemometers that can be used for this purpose. Be sure to measure only the headwind component of the wind by pointing the anemometer straight down the runway.

If a handheld anemometer is not available, it is possible to make use of the wind information as reported by the tower, FSS, or whatever facility is available. Winds as reported by such facilities will be given in the true direction from which they are blowing. It will be necessary to compute the headwind component of this wind if it is not very nearly directly down the runway. Headwind components can be calculated by use of a wind component chart shown in

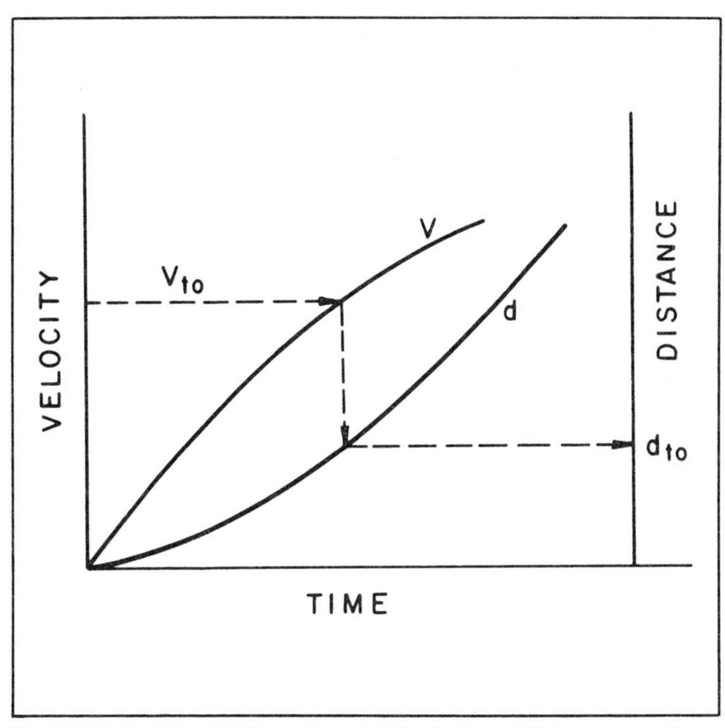

Fig. 5-4. Finding takeoff distance from plot of velocity and distance versus time.

Fig. 5-6. Before using this chart, you must first determine the angle between the wind and the runway. For a wind blowing from 270°, for example, and a runway numbered 24 (approximately 240°), the angle between the wind and the runway is 270° minus 240° or 30°.

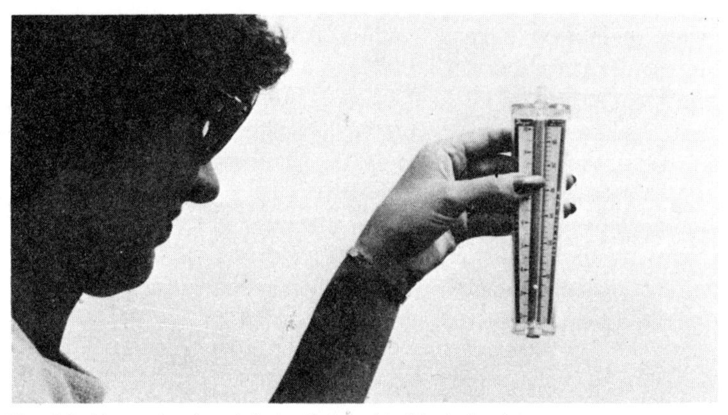

Fig. 5-5. Measuring headwind with hand-held windmeter.

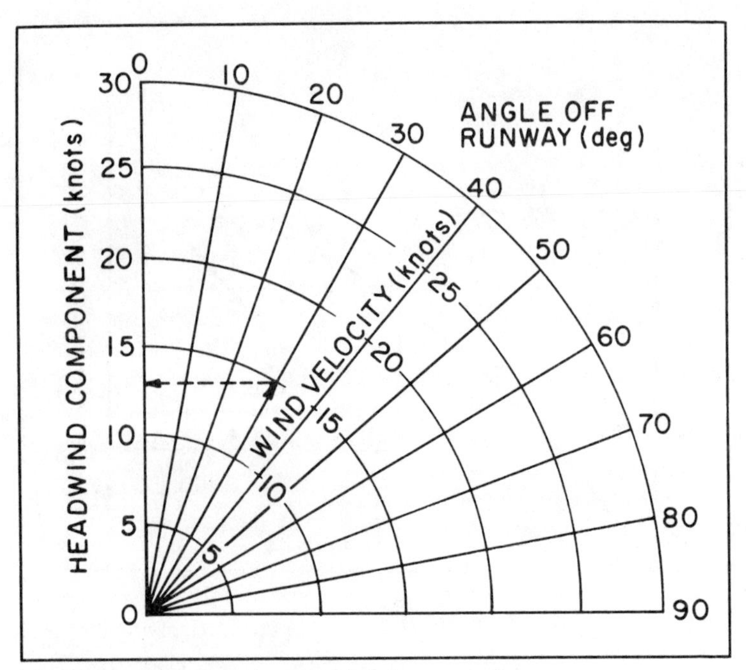

Fig. 5-6. Chart for determining headwind component.

Suppose a wind were blowing at this angle at 15 knots. From the chart, going out to 15 knots on the 30° line and then reading directly to the left, you would obtain a headwind component of 13 knots. Remember that wind is given in true direction and runway headings in magnetic. If much variation exists at your airport location it will have to be taken into account. In central Pennsylvania, for example, the variation is 8°W. 270° true direction would, therefore be 278° magnetic. In New England and the northwestern part of the United States, the variation can be 20° or more. Runway headings, of course, are rounded off to the nearest ten degrees. If you didn't know the exact direction of layout, just adding a zero to the number could result in an error of no more than plus or minus five degrees. The runway 24 used in the above example could actually head anywhere from 235° to 245°. This amount of error is negligible in calculating headwind components of wind.

The takeoff distance required under zero wind conditions can then be calculated from the following equation:

$$d_{o_{wind}} = \left(\frac{V_{to}}{V_{to} - V_{wind}} \right)^2 d_{wind}$$

The measured takeoff distance with wind, d_{wind}, is thus corrected by this factor using your calculated takeoff velocity, V_{to}, and the windspeed as measured, V_{wind}.

Altitude Correction

If the tests are conducted under conditions other than standard sea level (as they probably will be) a correction must be applied to yield the distance that would be required at sea level. Such corrections are not as exact as some relationships and various approximations have been devised. If the test condition is not too difficult from standard (that is, at a fairly low density altitude), a simple correction is possible using the density ratio. First find your actual density altitude from Fig. 1-3 and then the density ratio for this altitude from Fig. 1-4. The standard sea level takeoff distance is determined by multiplying the test distance by the square of the density ratio.

$$d_{std} = d_{test}\left(\frac{\rho}{\rho_o}\right)^2$$

The standard sea level, zero wind, value of takeoff distance can then be adjusted to other conditions by use of approximate correction factors. For airplanes with fixed-pitch propellers, the distance at any altitude (including fairly high density altitudes) can be approximated by dividing the standard sea level distance by the square and the square root of the density ratio for the desired altitude.

$$d_{alt} = \frac{d_{std}}{(\rho/\rho_o)^2\sqrt{\rho/\rho_o}} \quad (\rho/\rho_o \text{ for desired attitude})$$

This relation does not apply too well to constant-speed propellers. If you are testing an airplane so equipped, another method of approximating takeoff at altitude is by use of a Denalt computer. This is a device developed by the FAA to determine performance at various altitudes. For takeoff distance you set the appropriate air temperature and read opposite the desired pressure altitude a correction factor. Then simply multiply the sea level takeoff value by this factor to get the distance at altitude. There is one of these devices for constant-speed propellers and another one for fixed-pitch propellers. The Denalt computer for fixed-pitch propellers could also be used instead of the method given above. Figure 5-7 shows a Denalt computer.

It should be noted that these methods for obtaining takeoff performance at altitude by calculation from the sea level are only

Fig. 5-7. The Denalt Performance Computer used to determine take-off distance at altitude.

approximations. Experience has shown that the equation involving the density ratio sometimes yields results a bit short of the actual values as tested. On the other hand, the Denalt computer is a bit conservative and usually gives a little higher than actual. If you really want to know the takeoff distance at a fairly high density altitude (5000 feet or more), it would be wise to run actual tests at those altitudes. The procedure would be the same as the test outlined previously, except that it would be repeated at various density altitudes. Such a procedure would mean finding higher altitude airports or testing on higher density altitude days (usually meaning hotter days) or a combination of both.

The takeoff distance for any other weight can also be calculated by multiplying the standard distance by the square of the new weight desired divided by the square of the standard weight.

$$d_{\text{new wt.}} = d_{\text{std}} \left(\frac{W_{\text{new}}}{W_{\text{std}}} \right)^2$$

Starting with your standard sea level, zero wind, standard weight value, the takeoff distances for a number of other altitudes and several weights should be calculated. These values can then be plotted to give a workable takeoff performance chart as shown in Fig. 5-8. Note that these curves are almost straight lines but not quite.

Takeoff Distance to Clear Obstructions

Up to this point we have been considering takeoff distance as only that distance required to break ground. Another important consideration in takeoff is the ability to climb over obstacles that may exist near the end of the runway. Federal Aviation Regulations specify 50 feet as the standard height of an obstacle that must be cleared. In addition to ground run distance, therefore, you will usually find in airplane manuals the distance to take off and clear a 50 foot obstacle.

In determining total distance to clear a 50 foot obstacle, we usually break the takeoff distance into two parts. The first is the ground run distance, which we have previously been discussing. The second is the air distance. Air distance is the horizontal distance from the lift-off point to the point where the altitude is 50 feet. In other words, air distance is the portion of the total takeoff run during which the airplane is in the air.

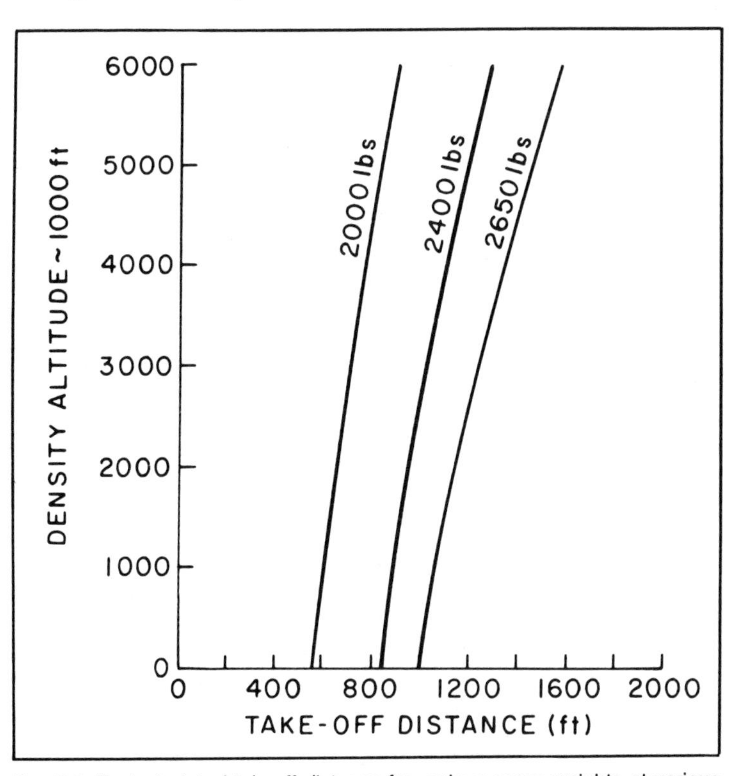

Fig. 5-8. Typical plot of takeoff distance for various gross weights at various density altitudes.

Measuring air distance, experimentally, is a bit complicated. For best results a vertically oriented radar is used. A movie camera photographing through a measured grid can also be used. Since most of us do not have such elaborate equipment, accurate location of the 50 foot altitude and the horizontal point at which it occurs is not easy. Even if it could be determined, the performance in this area is so dependent on pilot technique that many varying distances would result. A good determination depends on the pilot rotating at the correct speed to break ground and then attaining and holding the proper climb speed. FAR Part 23 states that at the 50 foot altitude the airplane must have reached 1.3 V_s or best angle-of-climb speed (V_x) plus 4 knots. All of these requirements impose considerable demands upon the pilot, especially when you consider that all of this has to be done within the first 50 feet of altitude. Therefore, even accurate measurements would not yield very reliable information unless quite a number of takeoffs were recorded and an average value determined.

A simpler method of determining air distance is to calculate it rather than measure it from tests (Fig. 5-9). The calculated air distance can then be added to the ground distance which was determined by tests. This can be done rather easily for airplanes with rather shallow climb angles, typical of most light planes. The assumption made is that the airplane lifts off at its best rate-of-climb speed. Rate-of-climb and best rate-of-climb speed will be determined in the next chapter. It will be seen that this best rate-of-climb speed (V_y) is very close to V_x plus 4 knots and also very close to our previously specified lift-off speed of 1.2 V_s. At shallow climb angles, you can assume that the horizontal speed of the airplane is approximately equal to its forward speed. Remember that in a climb the forward speed is in a direction somewhat inclined to the horizontal. With this assumption the air distance to reach a certain altitude can be calculated simply as the ratio of the horizontal to vertical speed times the altitude:

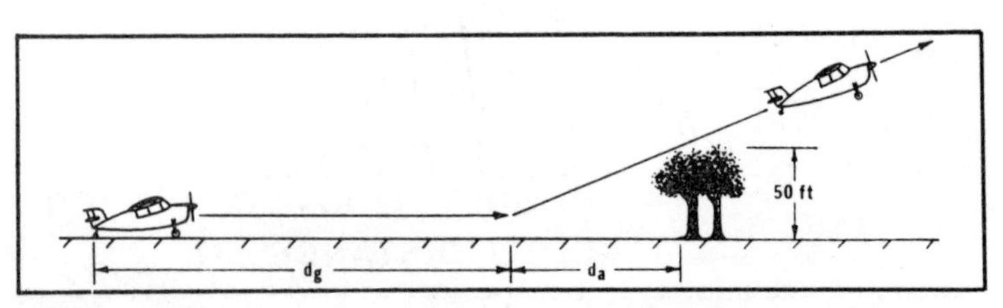

Fig. 5-9. Take-off over a 50 ft. obstacle showing ground distance and air distance.

$$d_A = \frac{V}{RC} \quad (50')$$

The velocity used can be takeoff velocity and the rate-of-climb value obtained from the tests in the next chapter. Use the rate-of-climb for the appropriate altitude. Since velocity is usually measured in knots or mph and rate-of-climb in feet-per-minute, some conversion factors must be included to make a workable equation. The results are as follows:

$$d_A = (4401)\,\frac{V_{t.o}}{RC} \text{ (for V in mph)}$$

$$d_A = (5061)\,\frac{V_{t.o}}{RC} \text{ (for V in knots)}$$

The total distance, then, to clear a 50 foot obstacle is the ground distance as previously determined plus the air distance:

$$d_{50\text{ ft.}} = d_G + d_A$$

Charts could also be made to include air distance and plotted similar to the ground run distance as in Fig. 5-7.

Takeoff Test Example

Test conditions:

- ☐ Gross Weight: 2200 lbs.
- ☐ Pressure Altitude: 2000 ft.
- ☐ Temperature: 38°F.
- ☐ Headwind (measured): 4 knots (5 mph).
- ☐ Distance vs. time as listed:

d(ft)	t(seconds)
100	5.5
200	8.0
300	10.0
400	11.5
600	14.0
800	16.5
1000	18.5

1. Plot distance versus time as in Fig. 5-10, and draw a smooth curve through the points.

2. Take time increments of 2 seconds and divide into corresponding increments of distance from the curve to calculate average velocities. Divide by 1.467 to get velocity in mph. Example:

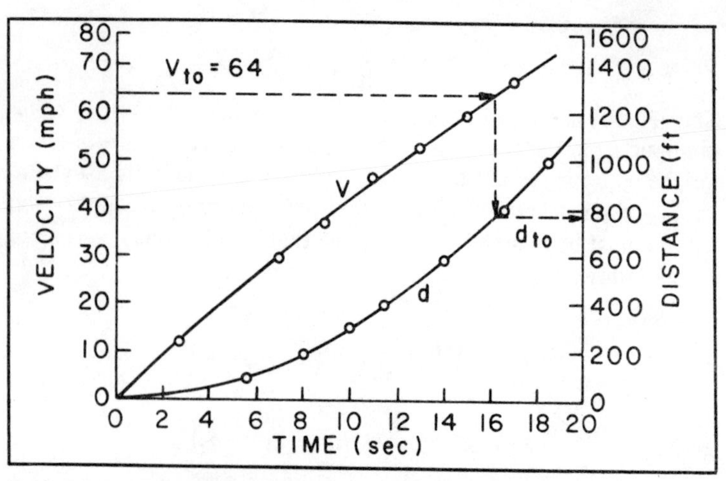

Fig. 5-10. Finding takeoff distance for the example given. For a takeoff velocity of 64 mph, the takeoff distance is 780 ft.

$$d_{av\,(4-6\,sec)} = \left[\frac{108' - 45'}{6\text{ sec} - 4\text{ sec}}\right]\left[\frac{1}{1.467}\right] = 21.5\text{ mph}$$

3. This point is plotted for the velocity at 5 seconds, the average between 4 and 5. Other average velocities are calculated similarly and plotted as the velocity curve in Fig. 5-10.

4. The stall speed for this configuration is established (by prior test) at 53 mph TAS. Takeoff velocity is calculated as 1.2 V_s.

$$V_{to} = 1.2(53) = 63.6\text{ mph}$$

5. This figure is rounded off to 64. Enter the chart plotted at this velocity, go straight across until the V curve is reached. At this point go vertically down to the d curve. Directly to the right of this point read 780 feet, the *test* takeoff distance.

6. This number is then corrected to zero wind from the test wind condition:

$$d_{o_{wind}} = \left[\frac{V_{to}}{V_{to} - V_w}\right]^2 d_{test} = \left[\frac{64}{64 - 5}\right]^2 780$$

$$= 1.177\ (780) = 918\text{ ft.}$$

7. This figure is further corrected to standard sea level. From pressure altitude of 2000 ft. and T = 38°F, by use of Fig. 1-3, density altitude is 1000 ft. From Fig. 1-4, the density ratio is 0.97:

$$d_{std} = \left[\frac{\rho}{\rho_o}\right]^2 d_{test} = (0.97)^2\ (918) = 864\text{ feet}$$

Takeoff distance for 2200 lbs, zero wind, and standard sea level is, thus, established as 863 feet.

8. Takeoff for other gross weights at sea level are calculated by:

$$d_{\text{new weight}} = 864 \left[\frac{W_{\text{new}}}{2200}\right]^2$$

Example:

$$d_{2000} = 864 \left[\frac{2000}{2200}\right]^2 = 714 \text{ ft.}$$

9. The distances for any of these weights at altitude are calculated by:

$$d_{\text{alt}} = \frac{d_{\text{S.L.}}}{(\rho/\rho_o)^2_{\text{alt}} \sqrt{\rho/\rho_o}}$$

Example:

$$d_{(3000 \text{ ft, } 2000 \text{ lbs})} = \frac{714}{(.915)^2 \sqrt{915}} = 892 \text{ ft.}$$

Chapter 6

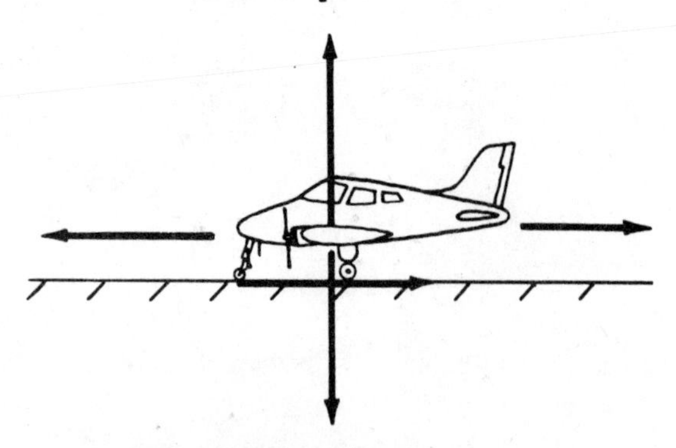

Climb Performance

The unique characteristic of an aircraft, compared to land-based vehicles, is that it can move in a third dimension. That dimension is vertical to the surface of the earth. The performance of the airplane in that dimension is, therefore, an important mark of the achievement of its intended purpose. Vertical motion away from the surface is referred to as *climb*. How well the airplane performs in this direction is usually determined by its rate-of-climb. The rate-of-climb depends both on the power output of the airplane's engine and on the drag of the airplane. In order to understand the behavior of the airplane in climbing or descending flight, it is first necessary to discuss both drag and power.

Drag

Aerodynamic drag is the resistance of the air which tends to retard the airplane in forward flight. Since forward flight is necessary to create lift, drag consequently plays a large role in climb performance.

Total drag depends on the configuration of the various airplane components and the way in which they are combined. The various ways in which the air creates drag on a component leads to several definitions for different kinds of drag. For a low speed airplane (below Mach .7) there are two major types of drag—parasite and induced.

Parasite drag is the resistance of any body to the air through which it is moving (or which is moving over it). It is generally broken down further into two parts: *form* drag (sometimes called *pressure* drag) and *skin friction* drag (sometimes called *viscous* drag).

The plate set perpendicular to the airstream as in Fig. 6-1A encounters practically all form drag, while the plate aligned with the airstream as in Fig. 6-1B encounters practically all skin friction drag. Since most bodies have a shape somewhere between these two extremes, such as the aerodynamic shape in Fig. 6-1C, they usually are subject to both types of drag. Thus, the term "parasite drag" is commonly used to denote the sum of both form and skin friction drag on a body.

Induced drag is an entirely different type of drag and applies only to the wing or other surface that creates lift. Lift is *defined* as the component of the aerodynamic force on the wing that is perpendicular to the velocity vector of the wind creating this force, such as in Fig. 6-2A. Because of the downwash, the wind vector is actually turned downward through a slight angle on an actual wing. The actual lift vector is therefore really perpendicular to this new velocity vector that results at the wing (Fig. 6-2B). In terms of our original definition of lift, that is, perpendicular to the velocity coming toward the airplane (or opposite to the flight path of the airplane), the lift is only the vertical component of the actual vector. Breaking up the lift in this manner also gives a component in the direction of the original velocity vector. Since forces in this direction are defined as drag forces, this component of the actual lift

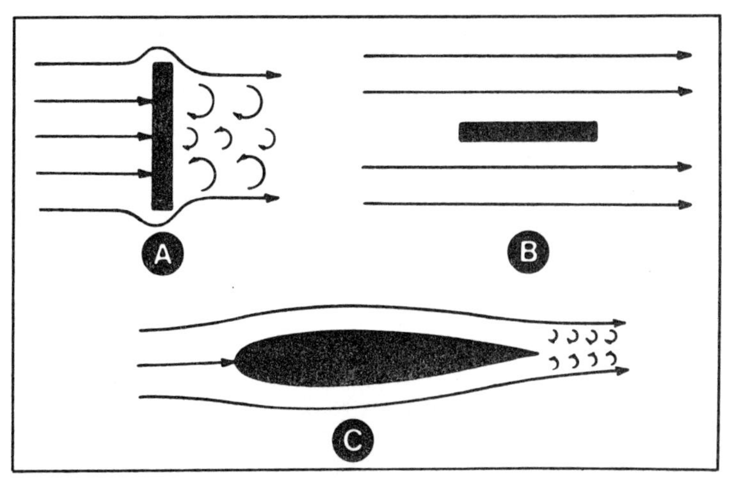

Fig. 6-1. Form and skin-friction drag. (A) Thin plate in this position encounters all form drag. (B) Plate in this position encounters all skin-friction drag. (C) Normal aerodynamic body has combination of form and skin-friction drag.

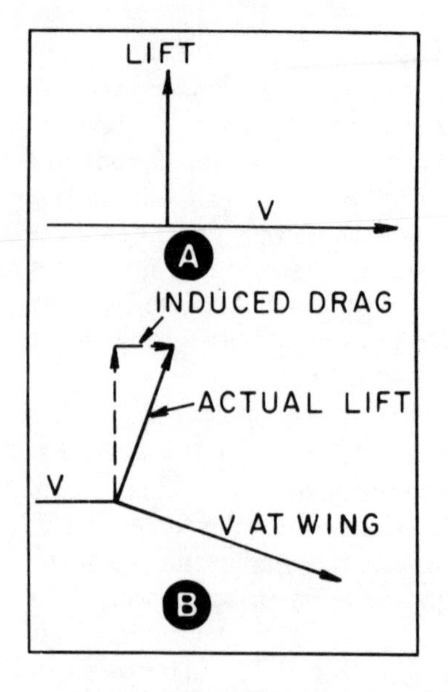

Fig. 6-2. Induced drag. (A) Lift is normally considered as force perpendicular to the airstream. (B) With downwash, the airstream is tilted downward at the wing, tilting lift vector backward, producing induced drag.

vector is termed "induced drag." It is an unavoidable consequence of the creation of lift by a wing.

Since parasite drag is caused by the dynamic pressure of the air, and dynamic pressure is proportional to the square of velocity (dynamic pressure = $\frac{1}{2} \rho V^2$), it follows that parasite drag is also proportional to the square of velocity. Induced drag, on the other hand, is proportional to the downwash and this increases at high angles of attack associated with low speeds. It actually works out that induced drag is *inversely* proportional to the square of velocity. In Fig. 6-3, the plot of drag versus velocity shows these opposite trends for parasite and induced drag. The total drag is the sum of parasite and induced and results in the characteristic curve as shown. This curve represents a plot of the drag equation:

$$D = D_{par} + D_{ind.}$$
$$= K_1 V^2 + \frac{K_2}{V_2}$$

Notice that total drag increases from some minimum point with *both* increase and decrease in velocity. This fact is very important to airplane performance, particularly to climb performance.

Power Required

Drag, then, represents the force that must be overcome by the thrust of the powerplant. Since the power of a reciprocating engine

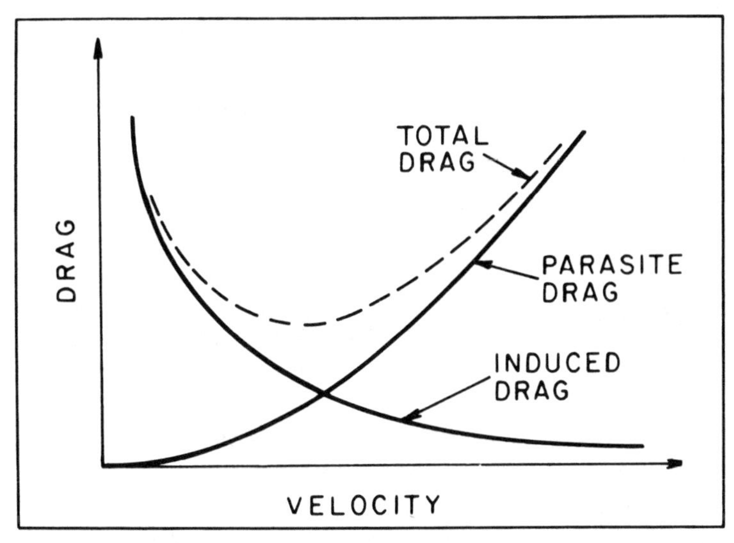

Fig. 6-3. Relationship of parasite drag, induced drag, and total drag with airspeed.

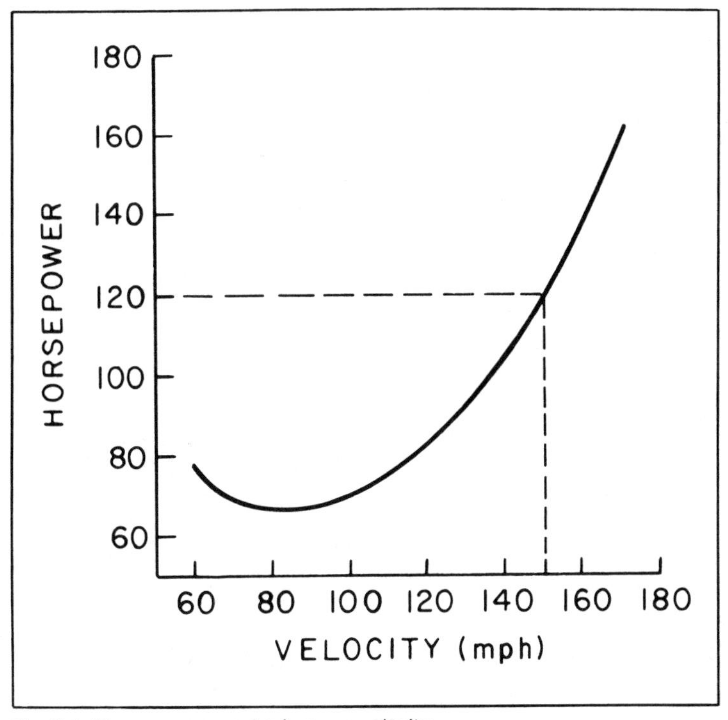

Fig. 6-4. Horsepower required versus velocity.

is easier to determine than the thrust of a propeller-engine combination, drag is often defined in terms of *power required*. Power is equal to a force times a velocity. Thus, if the drag force in the form of the above equation is multiplied by another velocity term, the power-required equation ends up as:

$$P_{required} = K_1V^3 + \frac{K_2}{V}$$

The first term is called the parasite power term and is proportional to the cube of velocity, while the second term represents the induced power and is inversely proportional to the first power of velocity. Power is usually expressed in units of horsepower and results in the same equation except that the constants (K_1 and K_2) have different values. When plotted, this equation yields a typical curve as in Fig. 6-4, commonly known among aviation in-groups simply as the *power curve*. This curve represents the power that is required at any velocity to maintain steady, level flight. For example, at 150 mph (knots if you prefer) this particular graph shows that 120 horsepower is required.

Power Available

Whether the airplane can actually fly at a certain value of power required (such as the 120 hp mentioned above) depends on whether that power is available from the powerplant. Obviously, if the engine is rated at 100 horsepower then it could never develop 120 hp. This would mean that the airplane could never fly straight and level at 150 mph in the example given.

The amount of power that can be obtained from the powerplant is termed *power available*. For any given engine, this power can be varied with throttle setting (and propeller pitch, if constant-speed). Also, the power at a given setting will decrease with altitude. Reciprocating engines display a relationship with altitude as shown in Fig. 6-5. This chart shows the horsepower at any altitude for a given rpm setting. The dotted line shows that at 5000 ft. and 2000 rpm the engine delivers 110 horsepower. This power is really the power available right at the crankshaft of the engine and is referred to as *brake horsepower* or bhp. If the engine is geared down to a lower speed at the propeller shaft, the power available there is called *shaft horsepower* or shp. The shp is a little less than bhp if gearing is employed due to frictional losses. For direct drive engines, however, shp is the same as bhp. Note that as altitude is increased at any specific value of rpm, the power drops off. This is because the

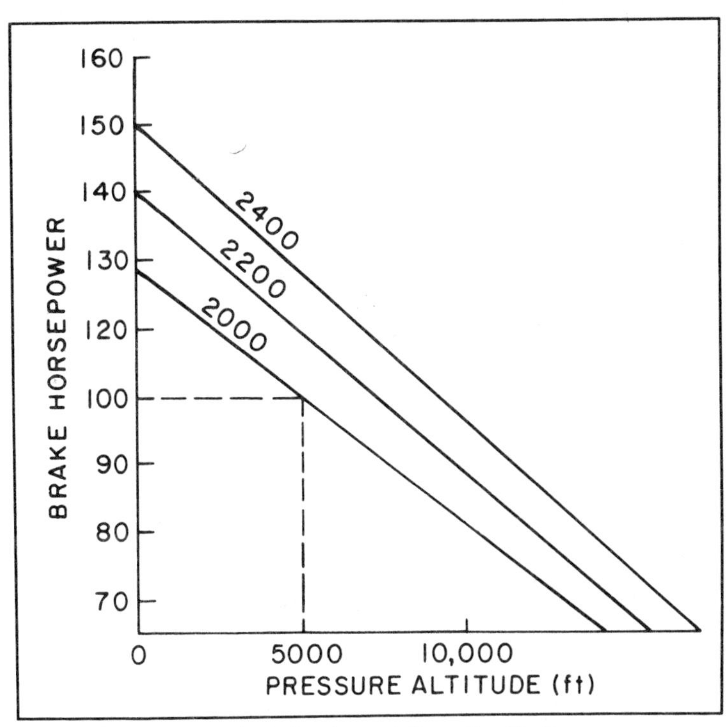

Fig. 6-5. Engine power chart showing variation with altitude.

density of the air gets less with altitude and the engine power output is proportional to the air density.

Regardless of the altitude, however, all of the bhp (or shp for geared engines) does not get converted into thrust. Thrust is provided by the propeller and a certain amount of power is lost in converting the torque of the engine into thrust force. The amount of power that does finally get converted into thrust is referred to as *thrust horsepower* or thp. It is this thp that must be counted on to match, or exceed, the power required due to drag.

The propeller acts basically like a wing in creating lift. Its "lift" vector however is pointed generally forward and the "lift" force that it creates thus becomes thrust. Since the propeller has an airfoil section and works like a wing, it also creates drag. Hence, some power is lost to the drag of the propeller. It works most efficiently at its maximum L/D (lift to drag) ratio. The propeller, however, is moving through the air at a forward velocity equal to that of the airplane, and also at some rotational velocity. A section of blade is shown in Fig. 6-6. Various combinations of forward and rotational

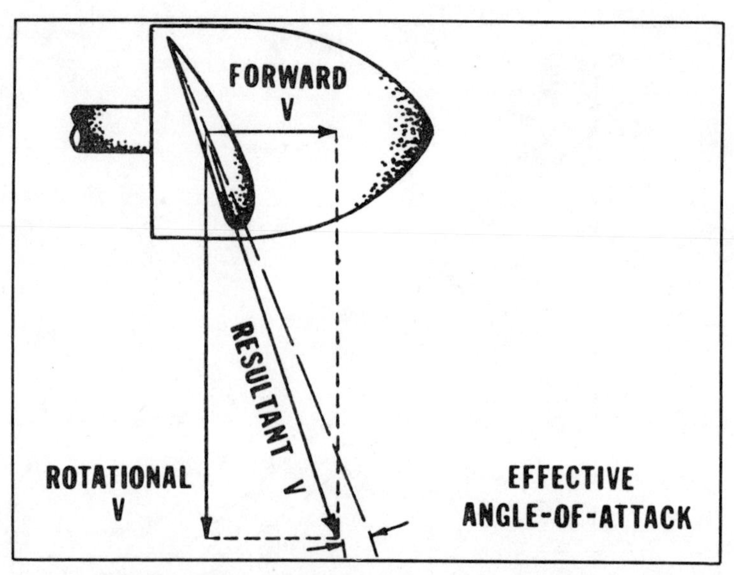

Fig. 6-6. Propeller motion showing forward velocity, rotational velocity, and resultant velocity from which angle-of-attack is determined.

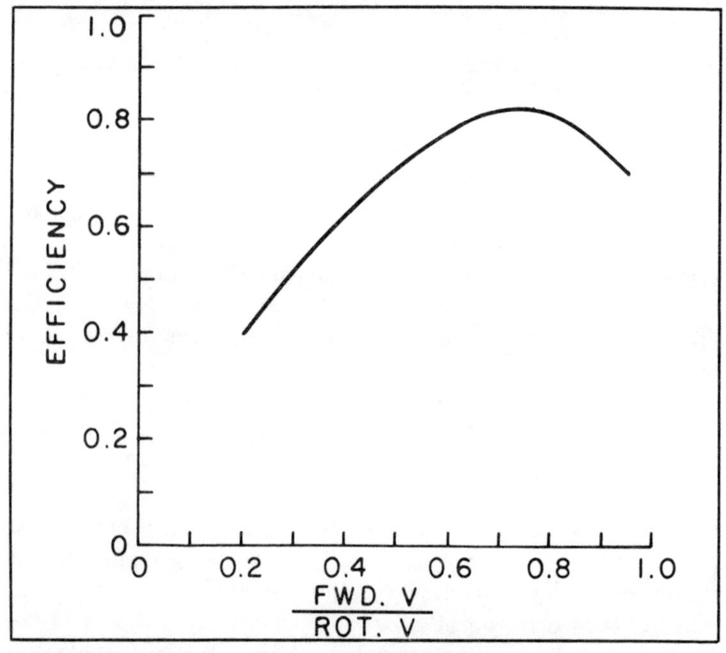

Fig. 6-7. Typical propeller efficiency curve as a function of forward to rotational speed ratio.

speed give different L/D ratios to the propeller and, consequently, different values of efficiency. A typical propeller efficiency chart is shown in Fig. 6-7. This chart shows that up to a point, higher forward speed at a fixed rpm value will increase propeller efficiency. If the propeller pitch can be varied, as with constant-speed propellers, the peak efficiency can be maintained over a longer range of forward speeds.

The horsepower available for thrust, or thp, is, then, the propeller efficiency multiplied by the bhp delivered by the engine. Since the bhp is a fixed value at a certain altitude and power setting, the thp varies with velocity proportionally to the way the propeller efficiency varies. Therefore, thp available when plotted against forward velocity has the same shape as the curve in Fig. 6-7. Remember that this curve represents power *available*. Such power is variable anywhere from zero up to the maximum capacity of the engine depending on throttle setting. Usually, therefore, we are interested in the *maximum* power available. This would be the curve of propeller efficiency multiplied by the maximum bhp for the particular altitude.

Rate-of-Climb

The airplane's ability to climb depends on the excess power available over that required. If the maximum power available is plotted on the same chart as the power required, the excess power available can easily be seen. Figure 6-8 shows such a plot. The excess power is the difference between maximum power available and power required at any given speed. The rate of climb is directly proportional to the excess power. Notice that in the figure, the *maximum* excess power is shown. This occurs at the velocity for best rate-of-climb, usually designated as V_y. At speeds both above and below this velocity, the rate-of-climb will decrease. Where the curves cross, the excess power becomes zero and climb is not possible beyond that speed. It is important, then, to know both the maximum rate-of-climb and the speed at which this rate occurs.

The curves shown are for only one altitude. At higher altitudes the maximum power available will decrease and the power required curve will be altered because of changes in drag. Therefore, the excess power becomes less with altitude and, hence, rate-of-climb becomes less. When there is no excess power at any speed, no further climb is possible and the airplane is said to have reached its *ceiling*. The altitude for absolute zero climb ability is called *absolute ceiling*. Another definition of ceiling is the *service ceiling*, which is a

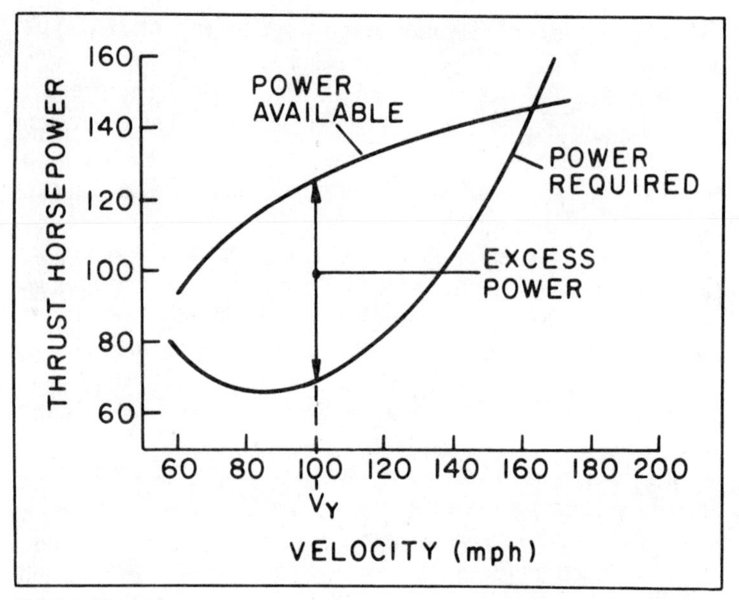

Fig. 6-8. Thrust horsepower versus velocity showing excess power as difference between power available and required.

point more practical to consider. This is the altitude where the climb ability is reduced to 100 feet per minute. One could conceivably fly at this altitude but not quite at the absolute value.

Test Procedure

In measuring rate-of-climb by flight tests you might be tempted to simply read the rate-of-climb (or vertical speed) indicator. However, this instrument is not always that accurate and usually has a great deal of lag associated with it. A timed measurement of definite altitude change is much better.

Speed or velocity is really the rate of time in which distance is covered. Average velocity is very easily determined by dividing a certain amount of distance by the time it takes to go that distance. Rate-of-climb is also velocity, but in a vertical direction. The average rate-of-climb can, therefore, be determined by climbing through a certain amount of altitude and timing the climb with a stopwatch. The rate-of-climb is, then, the altitude in feet divided by the time in minutes. The value will be for a certain airspeed which must be held during the measurement.

Before beginning the flight tests, make sure that the airplane is loaded to the standard weight (maximum gross or normal operating weight). If a standard weight below max gross is chosen, it would be wise to load to somewhat above the standard weight so that the fuel burnoff will reduce the weight to standard at about midway in the tests. In this way you will be operating closest to an average weight. Weight is very important to climb performance. You may even want to repeat the entire procedure for various average gross weights, since the rate-of-climb will be significantly greater at lower weights. On the other hand, the highest weight gives the minimum climb performance. Thus, if max gross or near max is tested, you can be assured that at lower weights it would not be any less.

Select a pressure altitude and fly to that altitude. If terrain permits you may want to start with an altitude of 2000 or 3000 feet. Begin the tests at a speed about 5 knots above stall or high enough so that buffeting or other indication of imminent stall is not encountered. Descend to 300 or 400 feet below the test altitude and establish a climb by applying full power and holding a constant airspeed. At 200 feet below the specified altitude start timing. Climb at a steady velocity until 200 feet above the test altitude and stop the timing at this point. Then descend again and repeat the procedure through the same altitudes at other airspeeds. Five or

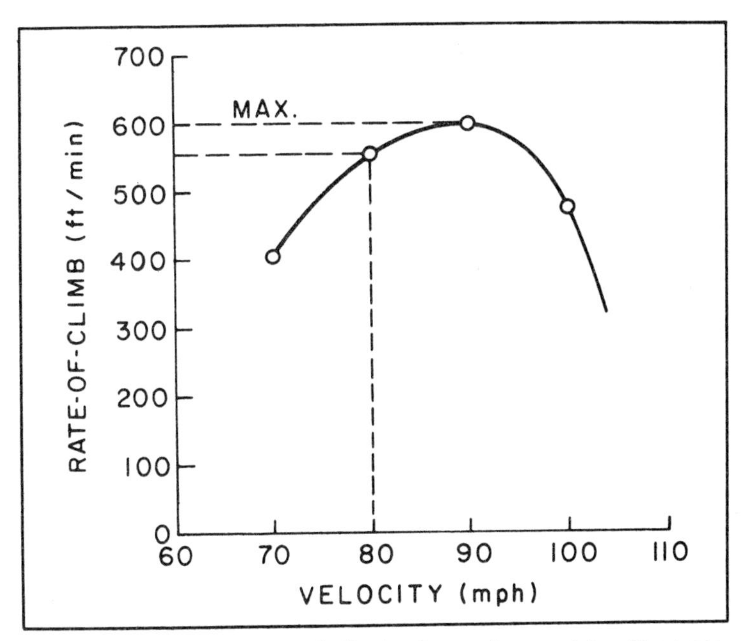

Fig. 6-9. Rate-of-climb versus velocity showing maximum point and best rate-of-climb speed.

ten knot (or mph) increments are sufficient. The times should decrease to a point and then start increasing. Once this trend is definitely established, the peak rate-of-climb is past and no further speeds at that altitude need be tested.

Determine rate-of-climb by dividing altitude change in feet by time in minutes for each speed tested. If time is measured in seconds, then multiply by 60 to get rate-of-climb in feet/minute.

$$\text{Rate-of-climb} = \frac{(\text{final altitude} - \text{initial altitude})}{\text{seconds}} (60)$$

Then plot the resulting rates-of-climb for the appropriate airspeed as shown in Fig. 6-9 and determine the maximum point by drawing a smooth curve through your test points. Note that you may not have measured a point right at the peak, but a smooth curve should indicate about where this peak should occur. Also, note the velocity

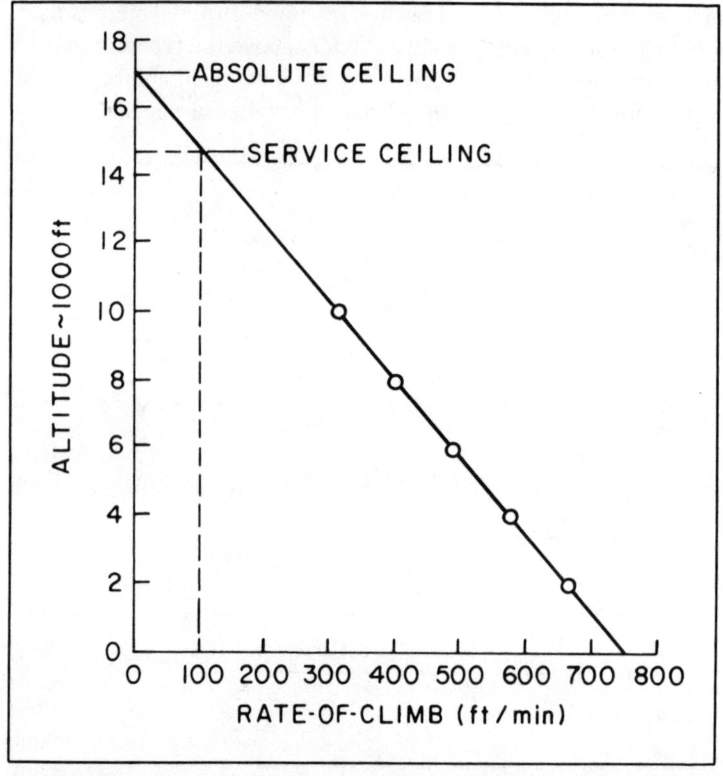

Fig. 6-10. Rate-of-climb versus density altitude showing determination of service and absolute ceiling.

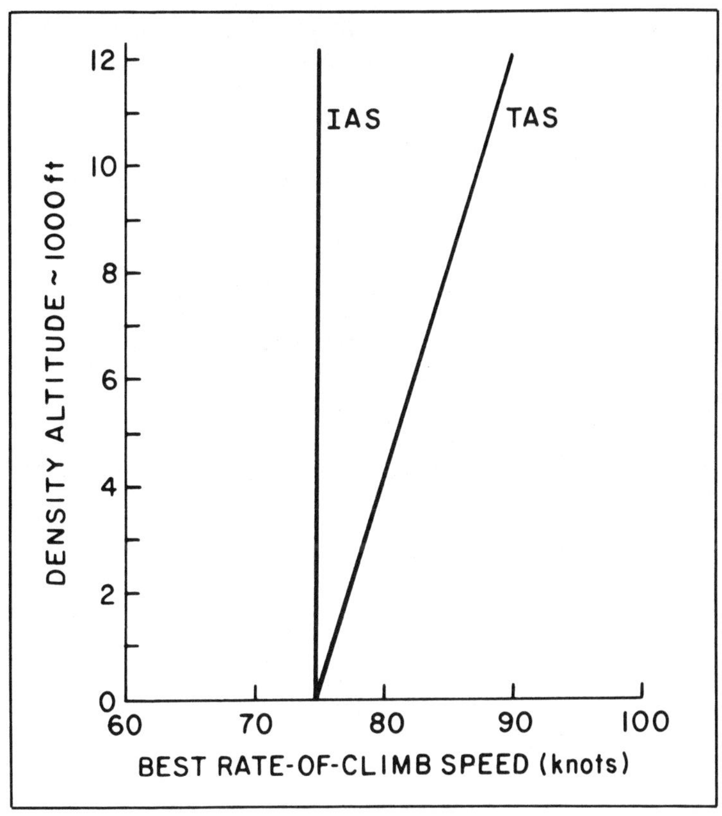

Fig. 6-11. Best rate-of-climb speed versus density altitude.

that falls directly under the peak. This is your best rate-of-climb speed, V_y.

Now, repeat this entire procedure for at least two other altitudes. Testing at a greater number of altitudes will result in more accurate final data. Then take the maximum values of rate-of-climb for each altitude and plot them against the respective altitudes as shown in Fig. 6-10. The altitudes, however, should be corrected to *density* altitude for this plot by one of the methods outlined in Chapter 1. The points should fall approximately on a straight line. Draw this line through the points and extend it until it touches both axes (zero lines) of your chart as shown.

The altitude where the rate-of-climb goes to zero is your absolute ceiling. The altitude corresponding to the rate-of-climb value of 100 ft/min is the service ceiling. The rate-of-climb for any density altitude can then be obtained directly from this chart.

Remember, though, to always calculate *density* altitude for your particular flight condition using the measured pressure altitude and temperature.

The best rate-of-climb speed for each altitude can also be plotted versus altitude. This speed should turn out to be nearly constant for all altitudes if measured in IAS or CAS. In terms of true airspeed, it will increase a bit with altitude. Such curves are shown in Fig. 6-11.

If the test airplane has retractable landing gear, you should measure rate-of-climb in both gear-up and gear-down configurations. The gear-down measurements need only be done at low altitudes to give an indication of initial climb ability right after takeoff. Twin-engine airplanes should also be tested for single-engine climb performance. Make sure that a competent multi-engine pilot is available for such tests and that all of the operating limitations for single-engine operation are observed.

Rate-of-Climb Test Example

Test Conditions:

- ☐ Test altitude = 5000 ft. (pressure)
- ☐ Initial altitude = 4800 ft. (pressure)
- ☐ Final altitude = 5200 ft. (pressure)
- ☐ Time = 43 sec.
- ☐ Temperature = 40°F.
- ☐ Airspeed = 80 mph IAS.

1. Determine rate-of-climb at this test point.

$$\text{Rate-of-climb} = \left[\frac{\text{final alt.} - \text{initial alt.}}{\text{time}}\right](60)$$

$$= \left[\frac{5200 - 4800}{43}\right](60) = 558 \text{ ft/min}$$

2. Plot this point on rate-of-climb versus velocity graph.

3. Assume other test points shown were obtained at this same altitude. Draw curve as shown. Maximum rate-of-climb is 600 ft/min.

4. From Fig. 1-3 correct 5000 ft. pressure altitude to density altitude:

Density alt. = 5500 ft.

5. Plot this point on altitude versus rate-of-climb chart.

Time and Distance to Climb

Another consideration in climb performance is the time and distance required to climb to cruise altitude. In wouldn't make much sense (or economy) to climb to an altitude that requires 20 minutes of climbing time, for example, if the overall flight only took 30 minutes. For short climbs, the time and distance could be calculated using an average value of rate-of-climb. However, when a climb to a fairly high altitude is planned, the rate-of-climb will change significantly from takeoff altitude to cruising level. In this case, using an average value of rate-of-climb would result in a significant error (Figs. 6-12, 6-13). For this reason it is best to actually measure the time to climb to various altitudes.

The test for time to climb is fairly simple. You take off and immediately establish best rate-of-climb speed and maximum recommended climb power. For most airplanes, maximum climb power is the full power available. Some, however, may have certain rpm or manifold pressure limits for continuous operation. Check the flight handbook or engine operating instructions for these limitations. Also, lean the engine according to the engine manufacturer's

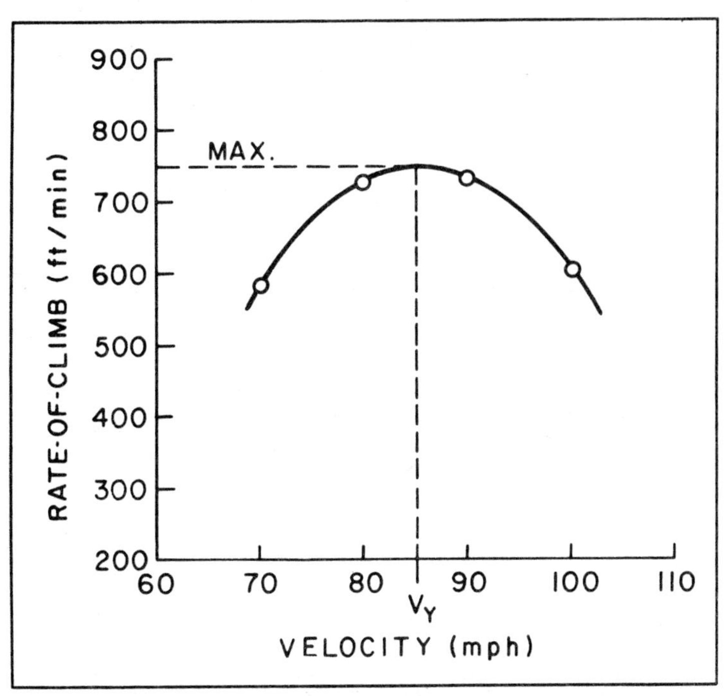

Fig. 6-12. Plotting rate-of-climb versus velocity and determining maximum.

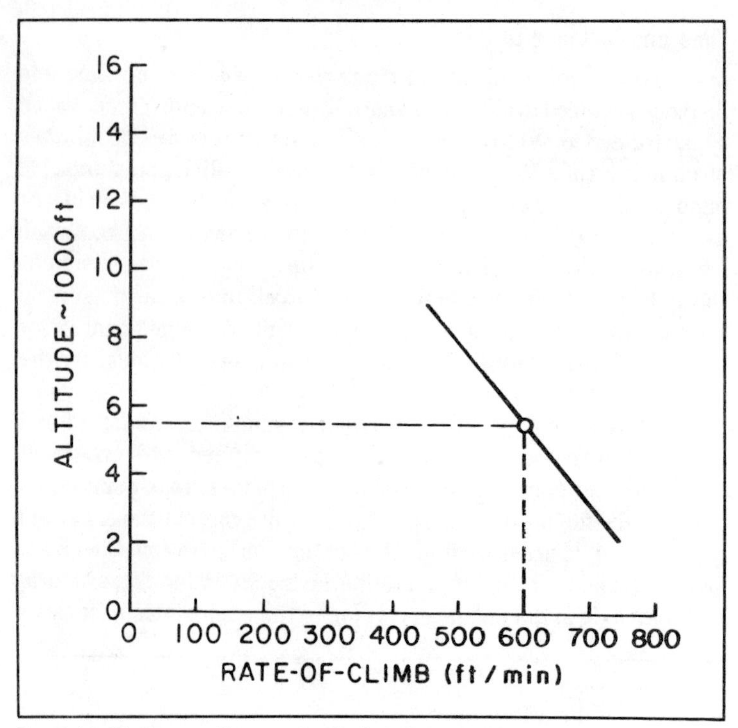

Fig. 6-13. Plotting maximum rate-of-climb for one altitude.

instructions. Many engines require full-rich settings in climb up to a certain altitude and then leaning to best-power setting can be done above that altitude.

Set the altimeter to 29.92 and time the climb to various pressure altitudes. Start the stopwatch at takeoff and leave it running. Read the altitude for every minute (or every 30 seconds) of elapsed time. Also, record the temperature at each reading and the total elapsed time. Continue this procedure up to, or close to, the maximum altitude at which you plan to fly. You may want to continue up to the ceiling. However, be sure to observe the special airspace restrictions that apply at high altitudes. VFR weather minimums increase above 10,000 feet and transponders are required above 12,500 feet MSL. Oxygen is also required above that level and recommended above 10,000 feet. For airplanes without such special equipment, it would be wise to limit climb tests to between 10,000 and 12,000 feet.

Using the pressure altitudes and corresponding temperatures recorded, convert the altitudes to density altitude values. Then simply plot the density altitudes reached against the respective times as shown in Fig. 6-14. One correction is necessary. Since your airport is probably not right at sea level, your climb will not start at zero altitude at time zero. This can easily be corrected for by extending the curve down to sea level. Since this point where the altitude is zero should occur at zero time, the whole curve can be shifted over to make it zero. The same amount of horizontal shift should be made to each point on the curve to yield a corrected curve, as shown. For example, if you took off at 1000 feet density altitude and the extended curve would cross the zero altitude line at −1 minute, this would indicate that one minute would have been required to climb from sea level up to the 1000 foot level. Therefore, you should shift all of the curve to the right one minute. In this way the zero altitude and zero time points will coincide, the 1000 ft. altitude will show +1 minute of climb time required, and so on.

To determine the horizontal distance covered during climb, the time-to-climb data can be used in a calculation method. The best rate-of-climb speed (V_y) in TAS at various altitudes will also be required for this calculation. This information can be obtained from a graph similar to Fig. 6-11 obtained from your rate-of-climb tests.

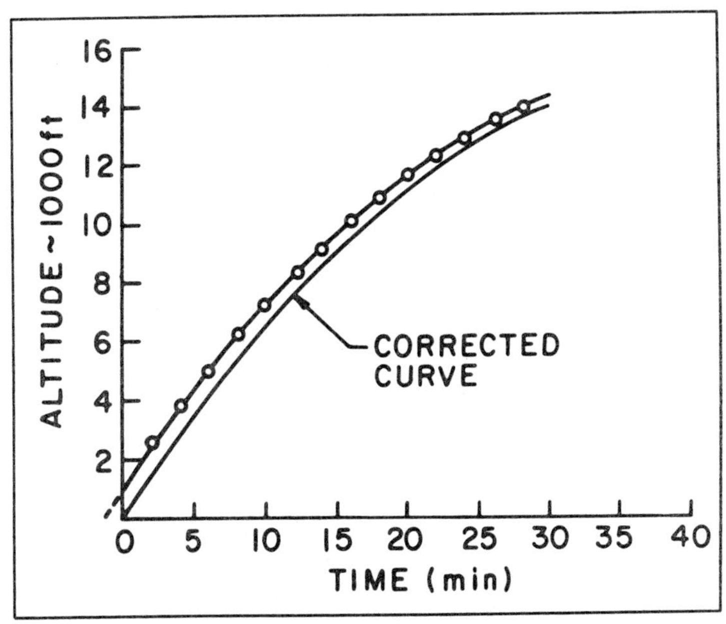

Fig. 6-14. Time to climb versus altitude showing correction for takeoff at standard sea level.

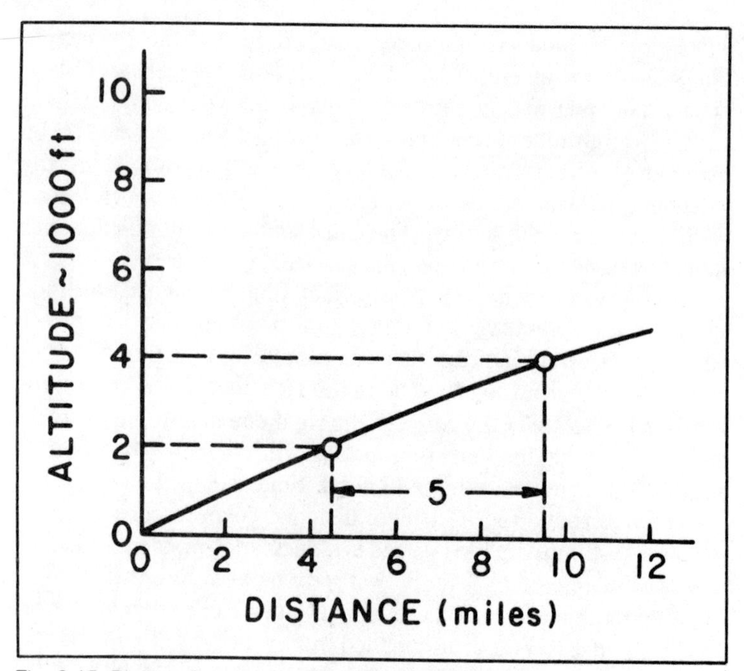

Fig. 6-15. Plotting distance to climb versus altitude.

Take the average V_y value between various altitudes and multiply it times the time required to climb between those altitudes:

$$d = V_{av} \frac{\Delta t}{60} \text{(for } \Delta t \text{ in minutes)}$$

As an example, assume the following data was obtained:

0 – 2000 feet: $\Delta t = 3.0$ min.
V_y @ 1000 ft = 90 kts
2000 – 4000 feet: $\Delta t = 3.3$ min.
V_y @ 3000 ft = 92 kts

$$d(0 - 2000) = 90 \frac{3.0}{60} = 4.5 \text{ nm}$$

$$d(2000 - 4000) = 92 \frac{3.3}{60} = 5.0 \text{ nm}$$

In this case, the 4.5 mile distance would be plotted at the 2000 ft. altitude point on the graph. The 5 miles would be added to this to yield 9.5 miles at the 4000 ft. point and so on as shown in Fig. 6-15.

Chapter 7

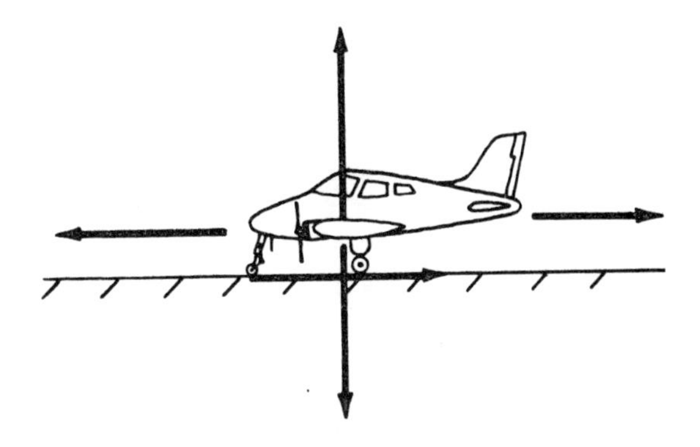

Cruise Performance

In the cruise situation, the pilot is usually interested in knowing how long it will take to reach the destination. This time depends on the groundspeed, which, in turn, depends on the cruising airspeed. In addition, the pilot needs to know whether he has sufficient fuel to reach the destination, or if not, how far the airplane can go before it needs to be refueled. Cruise performance determination thus means determining the cruising speed and the range associated with that speed. In certain operations, such as traffic patrol or other localized flight activities, the endurance of an airplane is significant. Endurance is simply the time that an airplane can remain aloft rather than the distance that it can go in that time. For most operations, however, it is *range* that is of more importance to the pilot.

Power Determination

Both the cruising speed and the range depend on the power output of the engine. More power creates more thrust, hence, higher speed. More power also means higher fuel consumption, however, and therefore usually results in reduced range. Before addressing the specific problems of speed and range determination, we should have an understanding of power and how to properly select it.

Reciprocating engines have a certain horsepower rating, which is usually the maximum power that the engine can produce at sea level when operated within design limitations. Engines are not usually run at maximum power continuously, however. Most light airplane engines are designed to run normally at 75 percent of the rated brake horsepower (bhp). Power is usually specified in this way, that is, as a percentage of the rated bhp. Thus, an engine rated at 200 horsepower would be delivering 150 horsepower at 75 percent bhp. The percentage applies to the rate horsepower and not the maximum power available at a certain altitude.

Of course, the engine can be run at less than 75 percent power and, consequently, consume less fuel. For most light planes, operation at less than 55 percent power results in very low cruise speeds. Thus, 55 percent is often regarded as the lower end of cruise power and is sometimes regarded as the *economy cruise* power setting. The upper limit, as mentioned, is usually 75 percent and referred to as *performance cruise* or *normal cruise* power. A good compromise, then, between the high fuel consumption at 75 percent and the low cruising speed at 55 percent is the middle range of 65 percent. Most airplane cruise performance charts will show curves for these three values of power. You can, of course, operate at points in between, or

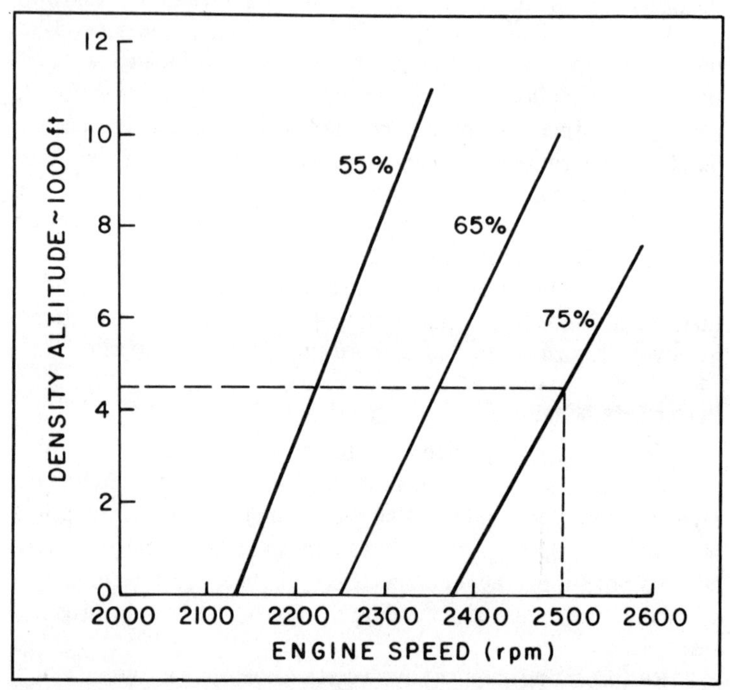

Fig. 7-1. Power chart for Piper Cherokee 140 showing RPM for various percent power at various altitudes (courtesy of Piper Aircraft Corp.)

even below 55 percent. The three values of 75 percent, 67 percent, and 55 percent merely represent degrees of performance that are typically, high speed, intermediate, and economy values, respectively.

Before proceeding to determine performance at certain power settings, we must know how to obtain a specific amount of power from the engine. For this purpose, it is necessary to consult the aircraft operating manual. If your test airplane is not a standard production model or has a modified engine installed, then it will be necessary to contact the engine manufacturer for power data. The manual for a production aircraft will provide power setting information such as is shown in Fig. 7-1. This chart is really for a Piper Cherokee 140 and shows, for example, that at sea level, 2375 rpm is the setting for 75 percent power. This airplane has a fixed pitch propeller which means that, at higher altitudes, a higher rpm is required to obtain the same amount of power, due to the lower air density. Therefore, at 4500 feet density altitude, a setting of 2500 rpm is necessary to obtain 75 percent power.

If the airplane has a constant speed propeller, the pilot has more of a choice in power settings. Such airplanes have both a propeller control (to set rpm) and a throttle (to set manifold pressure). Various combinations of these two controls can be used to obtain the same percentage of power. Table 7-1 is from the Piper Arrow handbook. It shows the combinations of rpm and MAP for various power settings and altitudes. For example, to obtain 65 percent power at 3000 feet, you could select either 25.1 inches of MAP and 2100 rpm or 22.2 inches and 2400 rpm. In either case, with the engine developing the same amount of power, the airspeed and the fuel consumption should be the same. It might be helpful, for flight test purposes, to plot such data in chart form if your handbook presents it in a table such as this one. In any case, you must have information which tells you what power setting you need to develop a certain bhp at a certain density altitude.

One other control which has a significant effect on the engine power output is the fuel/air mixture control. Performance that depends on engine power is usually indicated as being for a specific mixture setting. It is important for cruise performance that a certain mixture be maintained throughout the tests. This is best done by use of an exhaust gas temperature gauge (EGT). There are two mixture settings normally used in determining performance. One is known as "best economy mixture" and usually corresponds to the peak point on the EGT gauge. This mixture gives the lowest fuel

Table 7-1. Power Setting Data for the Piper Arrow II. (Courtesy Piper Aircraft Corp.)

Power Setting Table - Lycoming Model 10-360-C Series, 200 HP Engine							
Press. Alt Feet	Std. Alt Temp °F	110 HP - 55% Rated RPM AND MAN. PRESS. 2100	2400	130 HP - 65% Rated RPM AND MAN. PRESS. 2100	2400	150 HP - 75% Rated RPM AND MAN. PRESS. 2400	Press. Alt Feet
SL	59	22.9	20.4	25.9	22.9	25.5	SL
1,000	55	22.7	20.2	25.6	22.7	25.2	1,000
2,000	52	22.4	20.0	25.4	22.5	25.0	2,000
3,000	48	22.2	19.8	25.1	22.2	24.7	3,000
4,000	45	21.9	19.5	24.8	22.0	24.4	4,000
5,000	41	21.7	19.3	FT	21.7 FT	FT	5,000
6,000	38	21.4	19.1	--	21.5	--	6,000
7,000	34	21.2	18.9	--	21.3	--	7,000
8,000	31	21.0	18.7	--	21.0		8,000
9,000	27	FT	18.5	--	FT		9,000
10,000	23	--	18.3				10,000
11,000	19	--	18.1				11,000
12,000	16	--	17.8				12,000
13,000	12	--	17.6				13,000
14,000	9	--	FT				14,000

To maintain constant power, correct manifold pressure approximately 0.16″ Hg for each 10°F variation in inlet air temperature from standard altitude temperature. Add manifold pressure for air temperatures above standard; subtract for temperatures below standard.

consumption, but at a little sacrifice in power. The other setting is "best power mixture" which is the setting that gives the maximum engine power. It is usually achieved by backing off the mixture control to about 100° below the peak temperature. This setting can also be achieved without an EGT by leaning until a power loss is noted (sometimes accompanied by roughness) and then enriching until the engine just regains power and runs smoothly. This procedure, known as "normal leaning" is not recommended for fuel-injected engines. For any engine, use of the EGT gauge is best for flight testing.

The engine manufacturer's instructions should be followed closely in leaning procedures. Most engines should not be leaned at all above 75 percent power or above 65 percent in the case of geared engines. Since many engines will not develop 75 percent power above 5000 feet, flight manuals sometimes indicate not to lean below that altitude. You can really lean at any altitude as long as you know what you are doing. By that we mean knowing what your power setting actually is. If it is 75 percent or below for conventional engines, or 65 percent for geared engines, leaning can be accomplished safely.

Cruising Speed Test Procedure

Very basically, the cruise speed test simply involves flying at a certain power setting and reading the resulting airspeed. There are certain techniques that must be applied, however, in order to get accurate and meaningful data. Tests must be run at different altitudes to get the variation in airspeed with altitude. It is also desirable to run tests at various power settings, such as 55 percent, 65 percent, and 75 percent.

Begin the tests at as low an altitude as safe flight can be conducted. It would be desirable to start right at sea level if it were possible. Since this is not a very realistic situation, the lowest practical altitude can be used. It does not matter what exact altitude you select. You must, though, determine what the *density* altitude is at that particular true altitude. This is done by measuring temperature and pressure altitude and then using a calculator or the chart in Fig. 1-4 to get density altitude. Now you can go to your operator's manual to see what rpm (or rpm and MAP) you need to obtain certain power values at that density altitude.

Set up the power for 55 percent and fly straight and level. It is particularly important for this test to be done in very smooth air. Another trick in getting good speed readings is to fly for at least a minute, or better yet, several minutes at one setting. Make sure that no climb or descent tendency exists. When you are certain that the airplane is stabilized in level flight and the airspeed reading is holding constantly, take the airspeed reading. Then repeat the procedure for 65 percent and 75 percent, or other settings if desired. It would be advisable to run tests at least at 55 percent and 75 percent. The number of other settings chosen is up to you.

After completing tests for all settings at the first altitude, climb about 2000 feet and repeat the entire procedure. Do this for at least three different altitudes (or preferably, a few more). Eventually you will reach an altitude where you cannot obtain 75 percent power, but continue the tests at the lower power settings. If you go high enough, you will find that you will not be able to achieve 65 percent above a certain height. At this point you have probably gone high enough to obtain sufficient data for the entire test. Don't forget to always calculate the *density* altitude for each level tested.

Since cruising speed information will eventually be used to determine how fast the airplane is moving toward its destination, the speed values will have to be in *true* airspeed. Thus, it is necessary to convert all of the indicated airspeed readings to true. For this reason it is necessary to record the temperature and pressure altitude for each test point. Convert the IAS reading first to CAS by the calibration curve obtained from the test in Chapter 3. Then, correct CAS to TAS for the appropriate temperature and pressure altitude, either with a flight computer or by use of the density ratio and the equation in Chapter 3.

The resulting values of true airspeed can now be plotted against density altitude for the various power setting values tested, as shown in Fig. 7-2. The variation with altitude for a given power setting should follow a straight line. Some error in the testing is bound to occur, however, and the points will probably not fall exactly in line. Fit a straight line to the points as closely as possible. Also extend the line on down to sea level to determine what the cruise speed would be there. Do this for each power setting tested.

Maximum Speed Test Procedure

Although cruising is of most importance in cross-country flying, it is also desirable to know the maximum speed that could be attained at any altitude. Since maximum power is attainable at sea level, due to the highest air density here, the maximum possible level flight speed also occurs at sea level. The maximum speed then

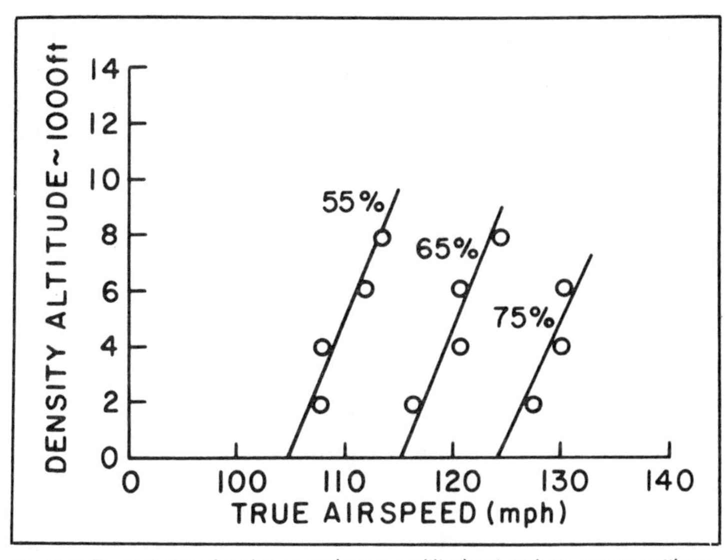

Fig. 7-2. Typical plot of cruise speed versus altitude at various power settings.

decreases with altitude as the density decreases. A plot of maximum speed against altitude is shown in Fig. 7-3.

Determining maximum speed is done by the same procedure as utilized for cruising speed, except that full power is applied

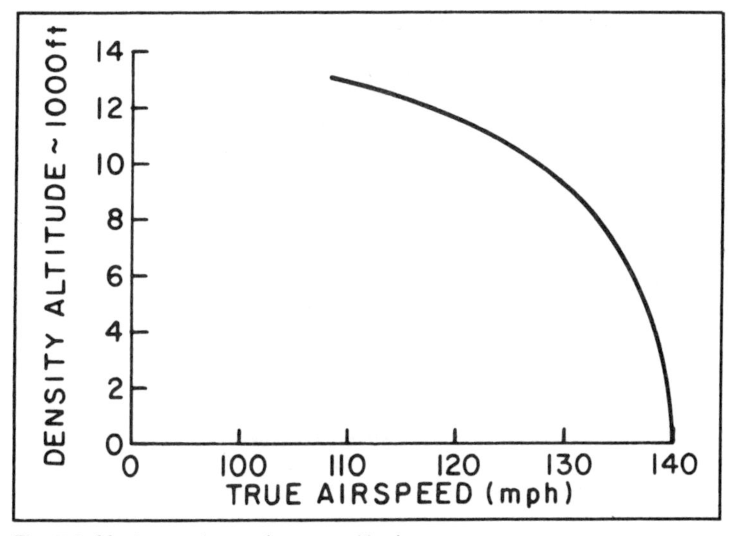

Fig. 7-3. Maximum airspeed versus altitude.

rather than some percentage of rated power. As before, it should be measured at various altitudes and the respective values of density altitude determined (Fig. 7-3). For maximum speed it is especially important to establish a true level flight condition and to hold that attitude for several minutes before taking the airspeed reading. Only the very slightest variation in angle-of-attack will cause a vertical movement and an erroneous maximum speed indication.

Caution should be used in operating at full power at lower altitudes. Most engines can be run continuously at their rated power without damage; however, engine wear may greatly increase above 75 percent. Other engines have a "takeoff power" rating which means that they can be operated at such power only for short durations of time, as would be done in takeoff. Most of the lower horsepower four and six cylinder engines can be run at maximum power indefinitely as long as engine temperature is within limits. If you are uncertain of the engine limitations, consult the manufacturer. If this is not possible, it is best to limit the maximum speed tests to higher altitudes where more than 75 percent power is not attainable. The approximate maximum altitude for 75 percent power should be apparent from the cruise power tests.

Determine the maximum speed at a number of density altitudes and convert the airspeed values to TAS. Then, plot these points on the cruise speed chart, as shown in Fig. 7-4. A curve fitted

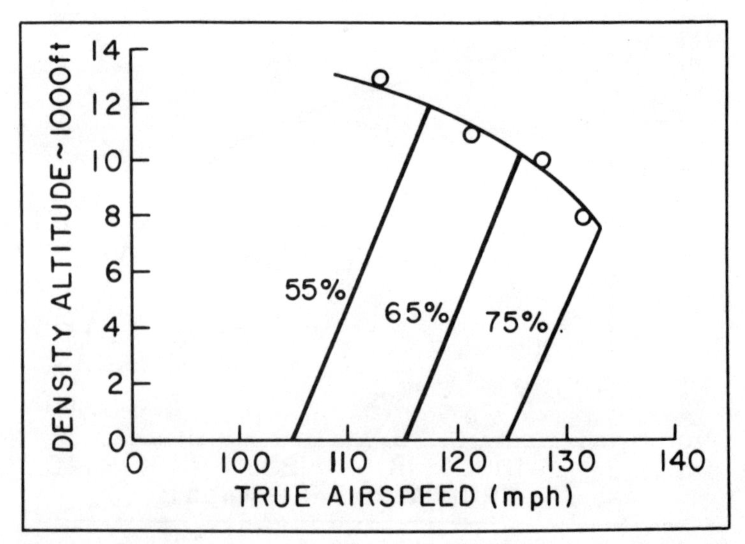

Fig. 7-4. Complete speed envelope constructed from cruise speed and maximum speed tests.

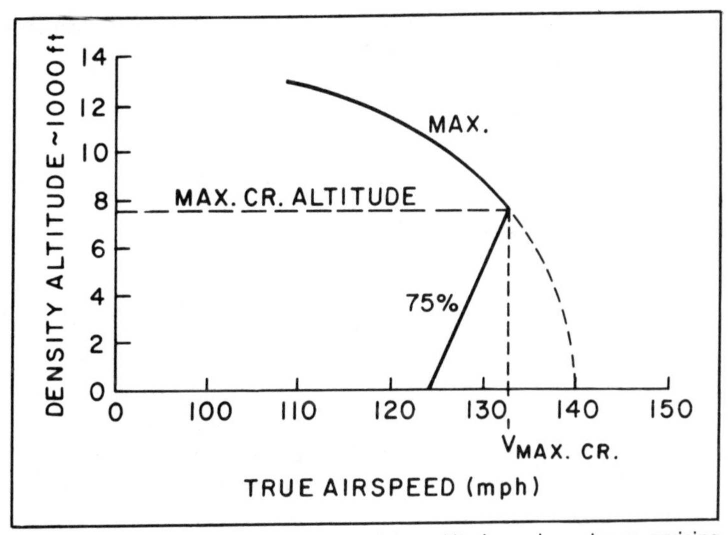

Fig. 7-5. Determination of optimum cruising altitude and maximum cruising speed.

to these points should then form the upper limit of your cruise speed lines.

Since 75 percent power is usually specified as maximum cruise power, the intersection of the 75 percent line with the maximum speed curve marks the overall maximum cruise speed point. The altitude for maximum cruising true airspeed. Above this altitude, the maximum *cruise* and the maximum (full power) speed are the same. Below this altitude, the maximum cruise speed is that for 75 percent power. This situation is apparent from Fig. 7-5. The extension of the maximum speed line below the max cruise altitude (shown as a dashed line) represents an area of *possible* speeds, but restricted due to engine design limitations.

Note that some airplanes with constant-speed props have a maximum recommended cruise rpm. The maximum practical speed for such aircraft would be determined by setting the prop control for this rpm value and then establishing full power by application of full throttle. Such procedure performed at various altitudes would then establish an upper limit on *cruising* airspeeds for a specified value of rpm.

Range

The range, or distance that an airplane can fly, depends on its cruising speed, its fuel capacity, and its rate of fuel consumption.

Specifically, it is determined as the speed times the fuel capacity divided by the fuel consumption. In equation form this is given as,

$$\text{Range} = \frac{V_{\text{cruise}}\ (\text{Fuel})}{(\text{fuel consumption})}$$

For range in miles, the speed would have to be in miles per hour, the fuel quantity in gallons (or pounds), and the consumption rate in gallons (or pounds) per hour.

Range, however, varies considerably for different power settings. We have already seen in the section on cruising speed how speed varies with different power settings at different altitudes. The fuel consumption also varies with power setting. In fact it is pretty much directly proportional to the bhp output of the engine. Since both speed and fuel consumption affect the range, it should be obvious that power has a great deal to do with determining range. At high power we get higher airspeed and this tends to increase range.

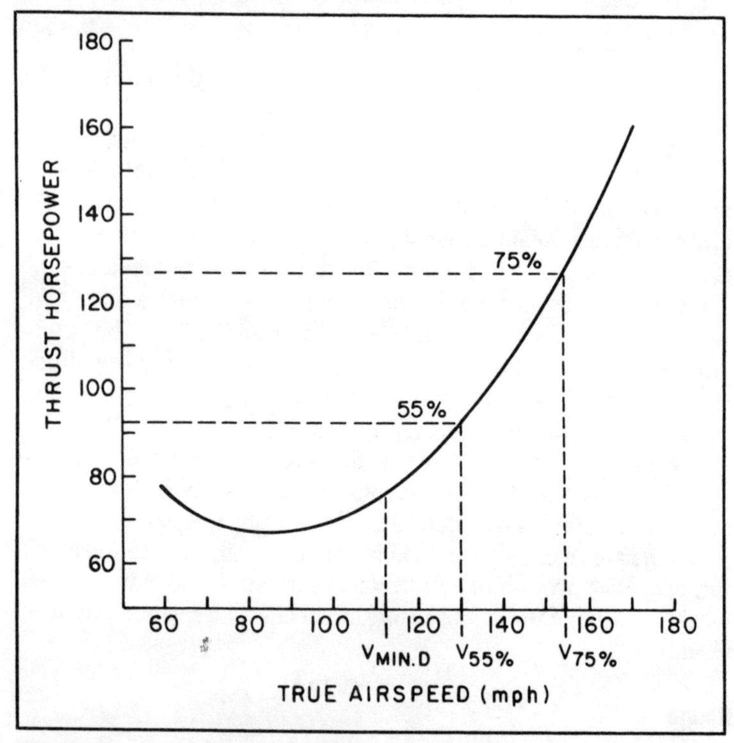

Fig. 7-6. Power required curve constructed from flight tests on Piper Arrow II showing power required at various level flight speeds.

However, high power requires high fuel consumption, and this term, being in the denominator of the range equation, will reduce range as it gets larger.

Figure 7-6 shows the power-required curve for a Piper Arrow II. This curve was constructed from flight test data. Notice that in the normal cruising range (55 percent to 75 percent power) that slower cruising speed requires a smaller amount of power. Less power is required because the drag drops off at lower speeds. Also, note that the power required drops off at a greater rate than the velocity in this area. Therefore, the overall result is an increase in range. This trend continues down to the point of minimum drag. Below this point, the velocity drops off at a greater rate than power, and range is *reduced*. Therefore, the point on the power curve where the drag is a minimum represents the power setting for *maximum* possible range.

Notice that the speed at this point of maximum range power is only 112 mph. This is pretty slow for an airplane capable of more than 150 mph cruise. We invest in airplanes with retractable gear and constant-speed props to attain some degree of cruise performance. At this rate you may as well have taken a trainer. The fact is that very seldom does anyone ever fly at maximum range speed. It is more of an academic consideration. We have already discussed, in the section on cruising speed, that most flying is done at power settings of between 55 percent and 75 percent bhp. In this power range, the Arrow cruises from about 130 mph to 153 mph. However, note that there is a considerable decrease in power required at 55 percent over that at 75 percent power. Thus, there would be a significant increase in range at the lower power setting.

Another quality of the airplane that affects range is the gross weight. The power curve, as shown, is really for one particular gross weight. At a lower gross weight the induced drag is less; therefore, a slightly higher speed is attainable at the same power setting. This fact complicates the exact determination of range, since weight is continually changing in flight. As fuel is consumed the airplane gets lighter. If fuel accounts for a major portion of the gross weight, the velocity corresponding to a certain power value could vary significantly. To calculate range, then, from the equation given above, you would have to use various values of velocity for various increments of weight to get increments of range, and then sum them for total range. For many airplanes, though, the fuel weight is a small portion of the gross weight and, in this case, an average value of velocity at average weight is sufficient. This is particularly true of many light single-engine airplanes whose total fuel weight is about 12 to 15 percent of the gross. For such airplanes, the velocity values obtained for various altitudes and power settings in the cruise speed tests are adequate to be used as constant values in calculating range.

Range Test Procedure

Just as the takeoff test does not require making an actual takeoff, range determination does not require flying the entire range of the airplane. Flying until the fuel is completely exhausted would be a bit foolhardy, anyway. Range is best determined by calculation from the equation given above. Testing is involved to determine the various terms in the equation. The amount of usable fuel should be known from the airplane operator's handbook. If the airplane is experimental, the actual capacity of the tanks can be measured when filling them. The cruise velocity for various altitudes and power settings should already be known from the cruise speed test.

The remaining item to determine is the fuel consumption. This determination will be the objective of the flight tests involving range. Fuel consumption rate is essentially constant for a given power setting. It could be measured at almost any altitude as long as the desired power is properly set. The power setting should be determined in flight just as it was for the cruise speed tests. Measure the pressure altitude and the temperature at your flight level and calculate density altitude either with a computer or the chart in Fig. 1-4. Then, from the airplane manual, obtain the power settings in rpm (or rpm and MAP for constant-speed prop) at that density altitude. Fly the airplane at the proper settings for 55 percent and 75 percent and any other settings for which data are desired.

An important step in setting up power to determine fuel consumption is to properly adjust the mixture control. Any desired mixture can be chosen for the tests, but then, it should be used consistently throughout. Most production airplane manuals use best economy mixture setting in determining range and best power mixture for cruising speed. Since cruise speed is measured with best power mixture, range could also be determined at this same setting. Best power mixture is also attainable without an EGT gauge (Fig. 7-7). Remember, though, that somewhat lower fuel consumption and slightly longer range could be achieved at best economy setting.

Fig. 7-7. Typical exhaust gas temperature gauge, very useful for determining mixture setting.

At each power setting, measure the fuel flow. This can be done in several ways. The best way is to temporarily install an accurately calibrated fuel flow meter in the fuel line. It will read directly and instantaneously in gallons per hour (or pounds per hour). Most airplanes with fuel-injected engines will have a fuel flow meter installed in the panel. Such a meter can be used for this purpose, but many of these do not score too highly in the accuracy column (Fig. 7-8).

Without an accurate fuel flow meter, another method has to be resorted to. This is to fly the airplane at a constant power setting for a significant amount of time and then measure the fuel necessary to refill the tank (Fig. 7-9). This procedure is fairly straightforward if your airplane has more than one fuel tank and each can be selected individually. In this case you make your takeoff, climb, descent, and landing on one tank and run on the other only when you are cruising with an accurate power setting. Keep the tank that is to be measured completely full until you have a definite amount of cruise power established. Then switch to that tank and fly for about an hour or two or whatever the tank capacity will permit. Before descent and landing, switch over to the original tank.

After landing, carefully monitor the refueling to determine the exact amount of fuel needed to replenish the cruise tank. Most flow

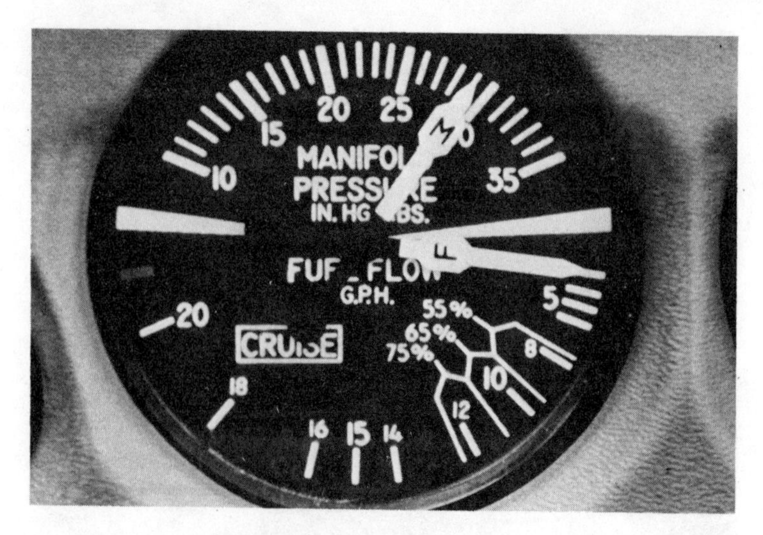

Fig. 7-8. Fuel consumption determination can be made directly from fuel flow meter.

indicators on fuel pumps and trucks are reasonably accurate. If you doubt the accuracy, you might have a gallon or two pumped into a measuring container and check it against the reading on the pump. A more exact way would be to pump all of the fuel into an accurate liquid measure and pour it from here into the tank. This is a bit time consuming and may try the patience of the line attendant. Use a bit of discretion in selecting an airport for your refueling. The FBO at Laguardia or Washington National may not be too cooperative in

Fig. 7-9. By measuring fuel consumed from fuel pump meter during refueling, consumption can be calculated.

this kind of procedure while he has three Learjets and a Sabreliner waiting to be refueled.

The fuel consumption can then be determined by dividing the numbers of gallons by the hours flown:

$$FC = \frac{\text{gallons}}{\text{time}}$$

Note that the consumption rate at only one power setting can be measured on each flight, if done in this manner. Another flight is required for each additional power value desired. Since several hours of flying time are involved in each of these tests you may as well combine it with a planned trip. It might even be a good excuse to make a trip. One power setting can be used on the way to the destination and another on the return trip. A three-legged course could also be planned for three power settings. It does not matter what cruise altitude you select, as long as you determine the settings necessary for a specific percentage of power at the altitude flown. The fuel consumption is essentially constant for a given power setting, regardless of other conditions.

If your airplane has only one fuel tank or, if individual tanks are not selectable, the procedure is a little more complicated. However, all is not lost. The test can still be done, although somewhat less accurately. The trick here is to use a given percentage of power for as much of the flight as possible. Switch to your desired power setting as soon as possible after takeoff and sufficient altitude have been achieved. Then retain that setting until just before landing. Remember that you can climb at 75 percent and other reduced power settings. You can also descend by increasing airspeed above that necessary for level flight. A long, shallow descent is probably the most difficult part of the maneuvers for this method. This can be done, however, right down to the traffic pattern. Again, choose your destination airport carefully. Don't plan such a test during an IFR flight into a TCA.

The important point in the single tank method is to keep the operations at other than cruise power settings to a small percentage of the overall flight. This procedure necessitates a fairly long flight. You must also have a pretty good idea of the power setting for your chosen power value right after takeoff. Then as you climb to a cruising altitude, you will have to continually reset your throttle for the same power at various altitudes. This is a somewhat demanding task and usually requires a crew of at least two. A preliminary flight could even be made to measure temperature and pressure at various altitudes. The exact setting for each altitude could then be determined and after filling the tanks, the actual test run could be made utilizing these settings. This variation of the basic method could be handled adequately by one person. Remember, also, to lean to the desired mixture setting at any altitude involved in the test.

Regardless of what exact method was used a fuel consumption figure for each power setting should now be established. Obtain cruise velocities for various power values and altitude from the chart constructed from the cruise speed tests. This chart should be similar to that shown in Fig. 7-4. Apply various values of speed and the proper fuel consumption value to the equation for range:

$$R = \frac{V_c(\text{fuel})}{FC}$$

It is best to read velocities from the lines on the speed chart rather than the exact measured points. The straight line curve in the speed chart tends to average out velocities and minimizes the error involved in measuring individual points on the curve.

Calculate the range for several altitudes for a given power setting (say 55 percent) and then repeat for another setting, such as 75 percent. Plot these points on a range versus altitude chart and draw a straight line through each set of points as shown in Fig. 7-10.

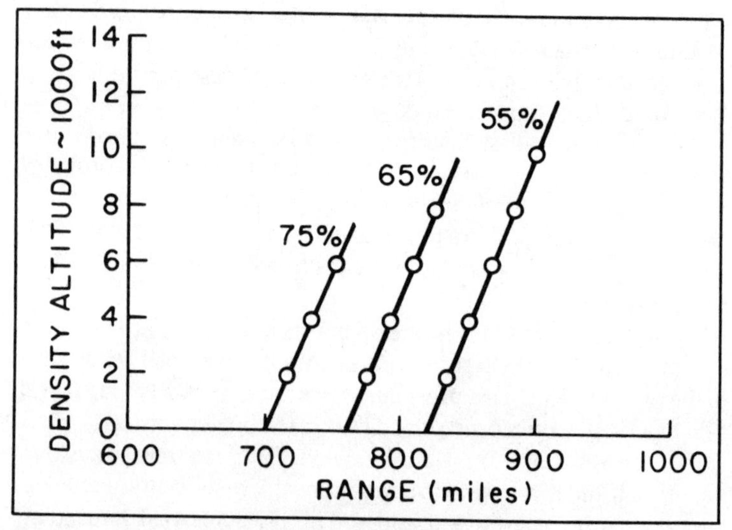

Fig. 7-10. Typical chart of range versus density altitude for various power settings as constructed from test data.

This chart now represents what is known as *absolute* range, if you have used the total fuel capacity in the equation. This range does not allow for climb and descent nor does it account for any amount of fuel reserve. Obviously, then, it does not represent an achievable range, but only a maximum limit. The actual distance you could safely fly (under no-wind conditions) would be somewhat less. Range charts often also include curves for range with a 45 minute reserve. The 45 minute reserve is determined as the distance that could be flown in 45 minutes at 55 percent power. This distance is easily determined by multiplying the speed at the proper altitude and 55 percent power by 0.75 (since 45 minutes is 0.75 hours):

$$d_{reserve} = 0.75\ V_{55\%}$$

This distance is then subtracted from your absolute range at that altitude to yield the range with 45 minute reserve. This same distance is subtracted from the 75 percent and 65 percent range curves as well, since the reserve distance is always considered to be flown at 55 percent power.

Endurance

For some operations it is desirable to know the amount of time that the airplane can remain airborne rather than the distance it can fly. This time is known as *endurance*. This figure is important, for example, if the pilot is in a holding pattern awaiting landing clearance. The endurance for a certain power setting is easily obtained by dividing the range at that setting and a certain altitude by the cruise speed at the same conditions:

$$E = \text{endurance} = \frac{R}{V_c}$$

Endurance is essentially constant for all altitudes. It also increases, like range, with decreased power. The maximum endurance is obtained at the very lowest point on the power curve. From Fig. 7-6, the speed at this power for the Piper Arrow is about 85 mph. Note that this speed is significantly lower than the maximum range speed.

Cruise Performance Example

Problem: Find cruise speed and range at 75 percent power.

Test condition: ☐ Pressure altitude = 5000 ft.
☐ Temperature = 55°F

1. Determine density altitude. From Fig. 1-3, density altitude is 6000 ft.

2. Refer to power chart for the airplane to get rpm for 75 percent power. Read 2620 rpm.

3. Set 2620 rpm, lean to best power mixture, and read velocity and fuel flow:

$$V_i = 110 \text{ kts}$$
$$FC = 10.25 \text{ gal/hr.}$$

4. Correct IAS to TAS. Assume negligible error in the airspeed system so that IAS = CAS. From Fig. 1-4, ρ/ρ_0 at 6000 ft. is 0.836:

$$\text{TAS} = \frac{1}{\sqrt{\rho/\rho_0}} (\text{CAS}) = \frac{1}{\sqrt{.836}} (110) = 120 \text{ kts}$$

5. Determine the range at this point (75 percent power and 6000 ft.). Fuel capacity is 48 gal:

$$R = \frac{V(\text{fuel})}{FC} = \frac{120(48)}{10.25} = 562 \text{ nm}$$

Chapter 8

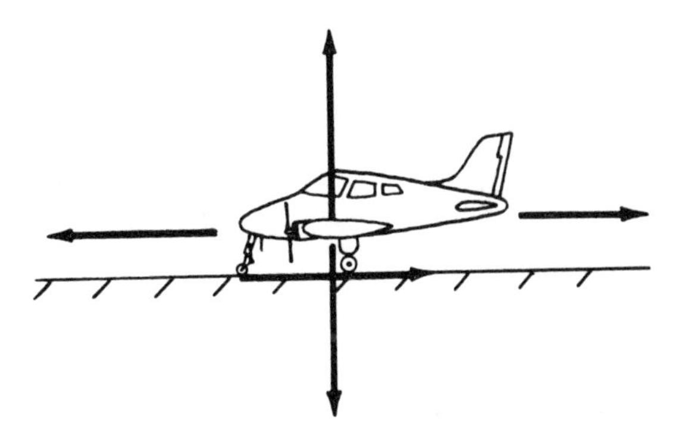

Descent and Landing Performance

In Chapter 6 we discussed how climb results when the power available is more than the power required. The rate-of-climb was shown to be proportional to the excess power available. If the power is reduced to below what is required at a particular airspeed, then a negative rate-of-climb value would result. Negative rate-of-climb is referred to as rate-of-sink, RS. The airplane would *descend* in such a condition proportional to the difference between power available and power required. Since the power available is less than that required, it could be referred to as a "decrement" in power rather than an excess, as in the climb case.

The maximum rate-of-sink, of course, occurs when the power is reduced to zero. The power available curve in Fig. 6-8 is shown at its maximum location and it was noted that the power controls could be adjusted to move this curve anywhere between maximum and zero power. If it were reduced to zero, then the power available would be the axis of the graph or line of zero thp. The rate-of-sink in this condition is now proportional to the difference between this axis and the power required curve at any velocity or to this decrement in power. Figure 8-1 shows this situation, which actually represents a power-off glide. Note that at the velocity for minimum power, the decrement of power would be smallest and, hence, the rate-of-sink would be lowest. As the speed increases or decreases from this point, the sink rate would increase. Gliding at the speed for minimum power would, thus, give you the longest *time* in a glide.

Power-off glide performance is of most concern in the event of an actual engine failure, in which case you really do have zero power available. If this situation should occur, it is usually the maximum glide *distance* that the pilot is interested in rather than glide time. The gliding distance possible from any given altitude depends on the ratio of the horizontal (or forward) speed to the vertical speed (or rate-of-sink). The higher this speed ratio the greater will be the gliding distance.

The maximum horizontal-to-vertical speed ratio could be determined if you had a plot of the power required curve. A line could be drawn from the origin (the point where both thp and V are zero) to where it just touches, or becomes tangent, to the power curve, as shown in Fig. 8-2. At the point where it just touches is the maximum *ratio* of V to RS. This ratio is also referred to as the glide ratio, GR. It turns out that glide ratio is the same as the lift-to-drag ratio (L/D). Therefore, this point of maximum horizontal-to-vertical speed is the maximum L/D point of operation and gives the maximum glide

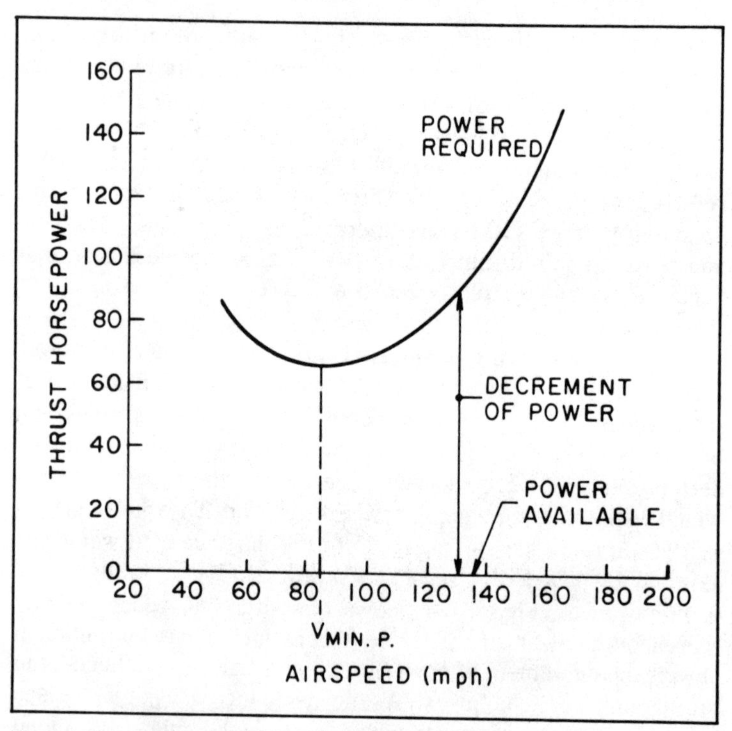

Fig. 8-1. Decrement of power resulting from zero power available as in power-off glide.

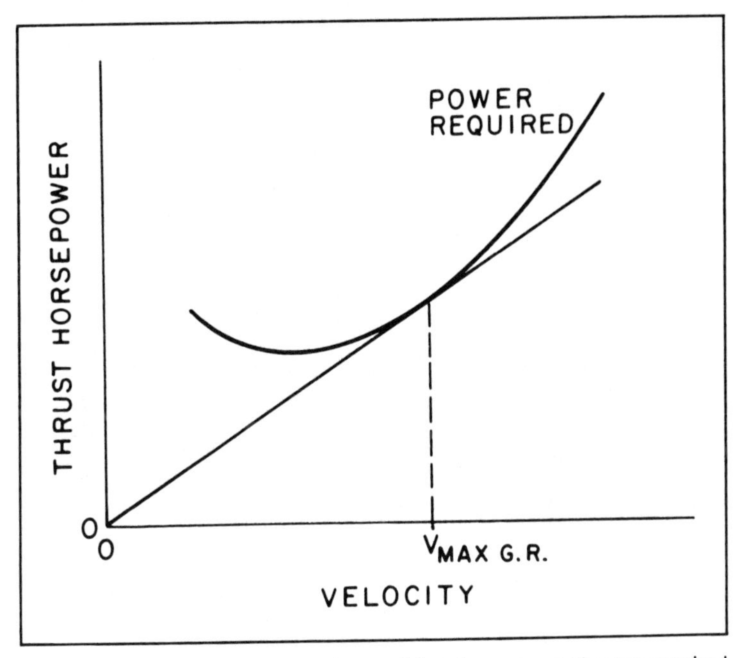

Fig. 8-2. Location of speed for maximum glide ratio on curve of power required versus velocity.

distance. You, of course, achieve this situation by flying at the velocity corresponding to this point on the velocity scale.

The distance, then, that one can glide is the glide ratio (or L/D ratio) at any particular speed being flown times the altitude (or vertical distance) from which the glide is started:

$$d_g = (L/D)\,h$$

The maximum glide distance would be achieved when flying at maximum L/D. Remember that the altitude here is the actual height above the surface and not above sea level. This relationship only holds for normal glides in which the glide angle is relatively small. In this case, the horizontal speed of the airplane is just about the same as the forward speed. This is not true in the case of a steep, screaming dive.

Determining Glide Ratio

The glide ratio is determined by measuring the rate-of-sink at various airspeeds. This is done much like the rate-of-climb determination except that a power-off glide is established rather than a full power climb. The time to descend through an increment of altitude is measured and rate-of-sink determined by dividing the altitude change by the time:

$$RS = \frac{\Delta h}{t}$$

This procedure is reported for various airspeeds and these speeds are recorded. The glide ratio is, then, the speed divided by the respective rate-of-sink at that speed:

$$GR = \frac{V}{RS}$$

Both horizontal and vertical speed must be in the same units. Rate-of-sink is probably measured most easily in feet per second. Therefore, forward speed will have to be converted to feet per second. If V is recorded in mph, the equation is:

$$GR = \frac{1.467\ (V)}{RS}$$

If V is recorded in knots, then the glide ratio equation is:

$$GR = \frac{1.687\ (V)}{RS}$$

The glide, or L/D, ratio does not change with altitude, so that it does not matter what altitude is selected for the experiments. The same velocity will also be associated with the same L/D as long as it is measured in *calibrated* airspeed (CAS). Velocity should be recorded in IAS and then corrected for airspeed error, if significant, to CAS. Weight does have a significant effect on L/D and, thus, careful attention should be given to loading the aircraft and recording accurate weight during testing.

Plot the resulting L/D ratios versus the calibrated airspeed to yield a curve such as shown in Fig. 8-3. The highest point on this curve will be your maximum L/D (or glide ratio) and the speed at this point will be your best glide speed. Using the equation given above for glide distance, the maximum distance for any altitude can be calculated by inserting your maximum L/D value:

$$d_g = (L/D)_{max} h$$

A plot of these calculations for several altitudes can be made as shown in Fig. 8-4. The glide distance could then readily be determined from this graph. Remember that you must fly at the speed for max L/D and close to the weight used in the testing.

Tests should be conducted with the engine idling. If a

constant-speed propeller is involved, it should be set at the highest pitch (full decrease). This gives the minimum propeller drag. In the event of an actual engine failure, the propeller should be adjusted to this setting. If the propeller were to stop completely, and not windmill, less drag would be encountered and the glide distance would increase somewhat.

To determine glide distance in landing configuration, the entire procedure should be repeated with landing gear down and flaps as normally set for landing. This test will also yield the L/D in landing configuration which is needed to determine the air distance portion of landing, as discussed in the following section.

Landing

Landing can be thought of as the takeoff process in reverse. In takeoff, the problem was to determine the distance to accelerate

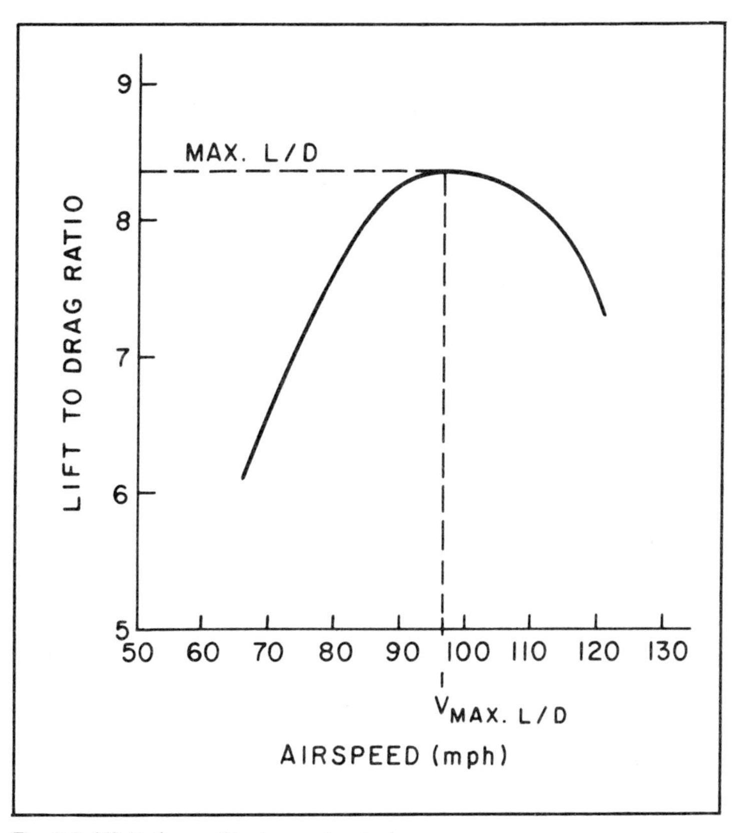

Fig. 8-3. Lift to drag ratio versus airspeed.

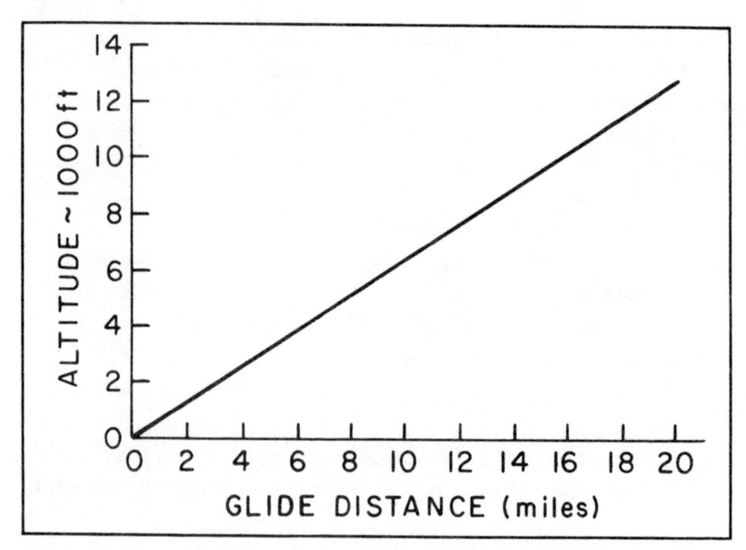

Fig. 8-4. Plot of glide distance versus altitude above ground.

from zero velocity up to takeoff velocity and then to climb over a 50 ft. obstacle. In landing, you need to know the distance to glide in over a 50 ft. obstacle, touch down, and then decelerate back to zero velocity. The distance is usually divided into an air distance and a ground distance, just as it was in determining takeoff performance. Some writers also consider a flare distance as a third part of the overall landing run. However, this distance can be included with the approach, or air distance, as we will do here. It should be realized that landing performance is highly dependent on pilot technique. There could be considerable variation in landing runs made by different pilots under the same conditions. A *possible* landing distance can be determined, however, just as a possible takeoff distance was considered.

Since the landing approach must be made as slowly as possible, but yet with a safe margin of speed, the stall speed plays a big role in determining landing performance. Federal Aviation Regulations require that landing distance for airplanes over 6000 lbs. gross weight be determined for an approach speed of not less than 30 percent above stall speed or 1.3 V_S. For lighter airplanes there is really no restriction on approach speed other than that it enables the pilot to make a safe landing. In any case, an approach speed of 1.3 V_S is a reasonable minimum approach speed. For an airplane that stalls at 60 mph, for example, this would mean a minimum approach speed of 78 mph, which is about as slow as most of us would want to fly it.

The regulations require (for those airplanes to which they apply) that the minimum approach speed be maintained down to a 50 foot height. Below that height, the speed can be further reduced. The air distance portion of the landing is thus the distance covered horizontally while the airplane is descending from 50 feet to touchdown, as shown in Fig. 8-5. If the airplane is considered to descend at a constant velocity of 1.3 V_s, then a flare distance is also required. If, on the other hand, the airplane is slowed down as the approach goes below 50 feet, the flare will be accounted for in the air distance (d_A). The more the airplane is slowed down during the air distance phase, the longer this phase will be. However, a slower touchdown speed results in a shorter groundrun phase (d_G). A full-stall landing, therefore, will result in the longest air distance but the shortest groundrun distance.

If the approach were to be made at a constant airspeed, the air distance could easily be determined by use of the glide distance equation given before. In this case the altitude is 50 feet and so the equation becomes:

$$d_A = \left(\frac{L}{D}\right) 50$$

All you would have to know is your L/D ratio; this ratio multiplied by 50 is your air distance portion of landing in feet. Landing at 1.3 V_s, however, is a bit hard on the shock absorbers, to say nothing of the occupants and the rest of the airplane. A gradual flare is usually made below 50 feet of altitude. If the airplane is flared to a full-stall at touchdown and the speed is 1.3 V_s at 50 feet, then the equation for air distance is:

$$d_A = (L/D)\,(50 + 0.0305\,V_s^2) \qquad (V_s \text{ in kts})$$

This equation is for stall speed (V_s) in knots. The 0.0305 V_s^2 part of the equation gives the additional distance required for flaring from 1.3 V_s to V_s. If V_s is known in mph, the equation is:

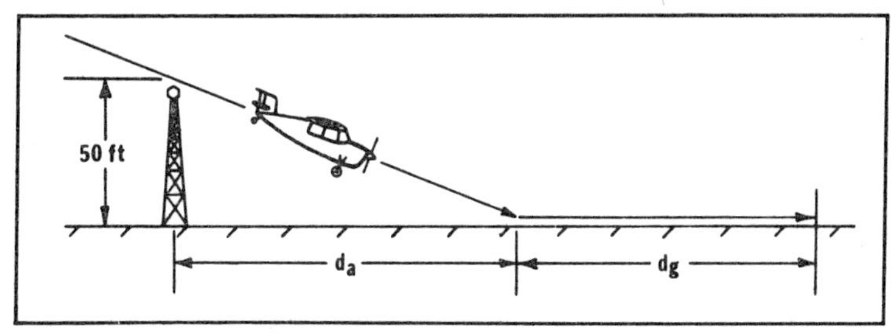

Fig. 8-5. Landing over 50 ft. obstacle showing both air and ground distance.

$$d_A = (L/D)(50 + 0.023\,V_s^2) \qquad (V_s \text{ in mph})$$

Many modern airplanes can be landed (and frequently are) at a speed somewhat above stall. A touchdown speed of 15 percent above stall is often used for calculating air distance when full-stall landings are not the normal procedure. In this case, the airplane is considered to be slowed from 1.3 V_s at the 50 foot point to 1.15 V_s at touchdown. For this situation, the above equations become:

$$d_A = (L/D)(50 + 0.01624\,V_s^2) \qquad (V_s \text{ in knots})$$
$$d_A = (L/D)(50 + 0.01228\,V_s^2) \qquad (V_s \text{ in mph})$$

The L/D ratio in these equations would actually be continually changing as the airplane is slowed down. An average value could be used, though, in calculating the air distance for landing. For the full-stall landing, the average velocity would be 15 percent above stall or 1.15 V_s and for landing at 1.15 V_s the average approach velocity would be 22.5 percent above stall or 1.225 V_s. The L/D should be evaluated at the appropriate one of these speeds. It should also be remembered that the L/D for use in landing distance determination must be measured in the landing configuration. Landing configurations, with landing gear and flaps down, are usually high-drag situations. Notice that the higher drag results in *lower* L/D ratios, which, in turn, results in shorter air distances for landing. This is the reason that flaps are normally used for landing. They steepen the approach by offering more drag and hence, shorten horizontal distances during the descent. Higher flap deflections result in progressively lower L/D ratios and subsequent shorter approach distances.

Ground roll is analogous to the takeoff run, except that the airplane must be *decelerated* from touchdown speed to zero velocity. The inertial force of the airplane is the only force that tends to keep it moving, as shown in Fig. 8-6, assuming that the power has been chopped completely. The drag and frictional force of the tires on the runway tend to slow it down. These retarding forces are rather small and, without additional help, would allow a pretty long landing roll. Indeed, if downward slope were present, the airplane could easily roll on forever (with enough runway) relying only on drag and normal rolling friction to stop. Fortunately, additional help is usually supplied in the form of brakes. Brakes greatly increase the frictional force and provide most airplanes with reasonably short landing ground rolls. Larger airplanes are also sometimes equipped with thrust reversers, deceleration chutes, or other special drag-inducing devices. For most light planes, however, the brakes are

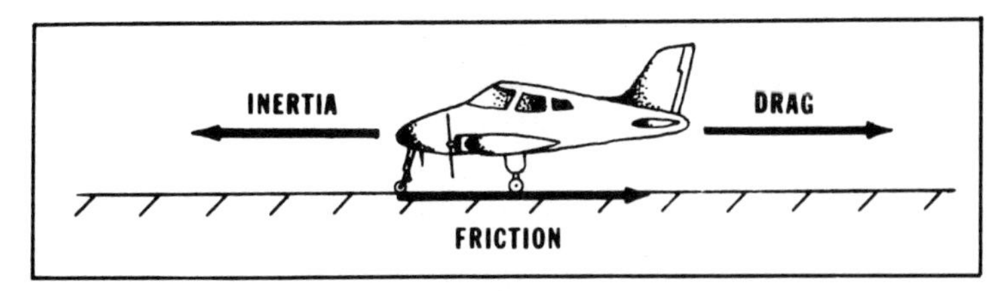

Fig. 8-6. Forces acting on airplane during landing roll.

the major source of decelerating power. They are expected to be used in determining the ground distance portion of the landing.

Landing Test Procedure

The experimental part of landing performance involves the determination of the ground roll distance. As mentioned, this distance is quite variable depending on pilot technique. It depends a great deal on the exact touchdown speed and on the amount and duration of braking during the ground roll. Even the same pilot will make different ground runs in repeated landings under the same conditions. For this reason it is necessary to make a number of trials and to take the average of these as the resultant distance. At least three, and perhaps as many as five or six runs will be necessary. It depends on how close the distances turn out to be on successive trials. If they are fairly close, only a few runs will be necessary. If the distances vary widely, then a larger number of trials will be required to yield a valid average figure.

The pilot should attempt to touch down at a definite speed. If he decides to obtain data for full stall landings, then the touchdown should be made in a complete stall. If landing distance is desired at a speed slightly above stall, the value of 1.15 V_s is recommended. Compute this value and attempt to touch down at that speed. For example if the stall speed in your landing configuration is 55 knots, 1.15 V_s would be 63 knots. Since the pilot will usually be sufficiently occupied in just making a smooth, precise landing, it might be wise to have an observer in the airplane monitor the airspeed indicator and note the speed at touchdown. Remember that the landing speed of 1.15 V_s is in TAS, so you should convert this to IAS and attempt to land at that value of airspeed, since you have to rely on the airspeed indicator to give you a reading. If the actual landing is made at a speed much different from the predetermined value, that run should be discarded and another run attempted.

During landing rollout, the pilot should apply a *moderate* amount of braking. Do not attempt to jam on the brakes immediately and with such force that the wheels skid. On the other hand, don't be too light on the brakes. Remember, that you are attempting to determine *normal* landing distance. Therefore, the brakes should be applied as in a normal landing procedure.

Spotting the actual touchdown point on the runway can be done by several observers on the ground similar to the way spot landing contests are judged. Touchdown zone stripes or fixed distance markers on the runway can be used as reference marks to locate the touchdown point. Runway lights could also be used, or for lack of any of these, stakes could be placed along the runway specifically for this purpose, as described in the takeoff test. Again, be sure to get the airport operator's permission to do this and make sure that they do not pose a safety hazard to normal airport operations. Also be sure that observers stationed along the runway are familiar with aircraft operations and remain at a safe distance from the runway. An approximate touchdown point could also be determined by an observer in the airplane using the same sort of reference markers (Fig. 8-7).

Bring the airplane to a complete stop and, again, note the exact stopping point. This too can be done best by a ground-based observer, but is possible from the airplane. The distance, then, is determined by measuring between the touchdown point and the stopping point. This measurement could be done most accurately by use of a tape measure. However, if runway lights or other distance markers of known spacing are used, the distance could be estimated by reference to them. The points would have to be spotted as approximately ½, ⅓, or whatever fraction, of the distance between

Fig. 8-7. Runway markings which can be used to determine landing touchdown point.

them. Remember that landing distance can't be determined down to the exact foot. A length to the nearest ten feet is probably the best that can be expected.

The tests should be conducted in calm wind, on a level runway, and at a gross weight close to the selected standard weight. Of course, the tests should also be run on the type of runway surface for which you desire the distance information. If the conditions are not exactly standard, some corrections can be made. A slight wind can be compensated for in the same way that it was in the takeoff performance. The distance measured in a slight headwind can be corrected to what it would be in zero wind by the following equation:

$$d_o = d_w \left[\frac{V_L}{V_L - V_w} \right]^2$$

V_L in this equation is the true airspeed at landing and V_w is the headwind component of the windspeed. The term d_w is the distance as measured with a headwind equal to V_w. The resulting distance for zero wind (d_o) can then further be corrected to standard sea level by multiplying by the density ratio at the test condition:

$$d_{SSL} = d_o \left[\frac{\rho}{\rho_o} \right]$$

If the weight at the test condition is different from standard weight, another term in the above equation would correct to standard weight. With the weight correction term added the complete equation is:

$$d_{SSL} = d_o \left[\frac{\rho}{\rho_o} \right] \left[\frac{W_{std}}{W_{test}} \right]^2$$

Landing ground roll distance for other altitudes and other gross weights can be calculated from the standard sea level value just as it was for takeoff ground run. The distance for higher altitudes is obtained from an equation involving the density ratio:

$$d_{alt} = \frac{d_{SSL}}{(\rho/\rho_o)_{alt}}$$

Simply divide the standard sea level value by the density ratio at the appropriate altitude. Landing roll for a lower gross weight can also be determined by first correcting the standard sea level value at standard gross weight to the new gross weight:

$$d_{new\ wt.} = d_{std.\ wt.} \left[\frac{W_{new}}{W_{std}} \right]^2$$

The distance for this new weight can also be computed for higher altitudes by using the equation given above for altitude correction, except that you start with the distance for this new weight ($d_{new\ wt.}$) in place of d_{SSL}. The resulting data can be plotted on a working chart to be used under any weight and altitude conditions as shown in Fig. 8-8.

The above discussion all involved only the ground roll distance (d_G). The total distance to clear a 50 foot obstacle would have to be computed by adding the air distance to the ground roll distance. The air distance would be calculated by use of the appropriate equation given in the previous section. The proper equation depends on whether a full-stall or above-stall landing is made and whether stall speed is known in knots or mph. For a full-stall touchdown and V_s in knots, for example, the air distance is calculated as:

$$d_A = (L/D)\ (50 + 0.0305\ V_s^2)$$

The L/D ratio would have to be determined experimentally as outlined in the section on glide ratio test procedure. For full-stall landing, it should be measured at 1.15 V_s.

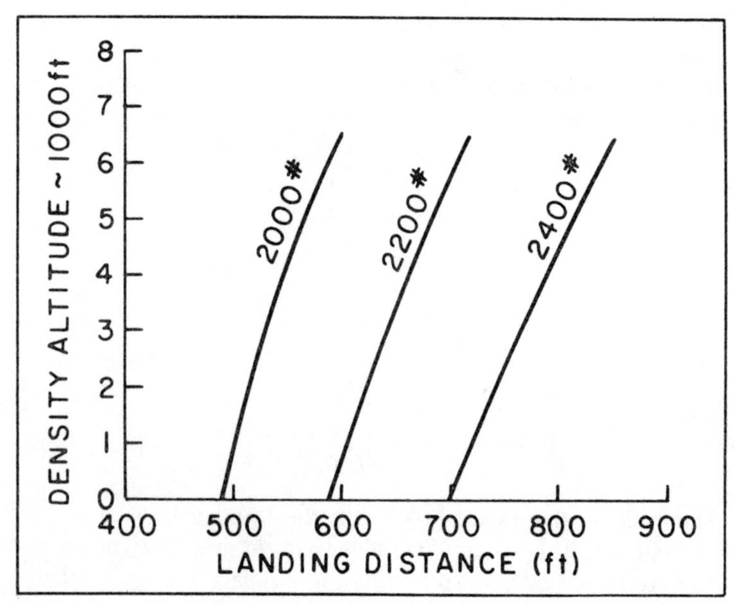

Fig. 8-8. Plot of landing distance versus altitude for various gross weights.

Glide Distance Example

Test Conditions:

- ☐ Altitude change = 400 ft.
- ☐ Time = 27.3 sec.
- ☐ Indicated airspeed = 100 mph

1. First, calculate rate-of-sink.

$$R/S = \frac{\Delta h}{t} = \frac{400}{27.3} = 14.65 \text{ ft/sec}$$

2. Now, determine L/D ratio.

$$L/D = \frac{1.467\ V}{R/S} = \frac{1.467(100)}{14.65} = 10$$

3. Plot this point against velocity as shown in Fig. 8-9.

4. Assuming similar calculations for other velocities have yielded the curve as shown, determine $(L/D)_{max}$ as the highest point on this curve.

$$(L/D)_{max} = 10.4$$

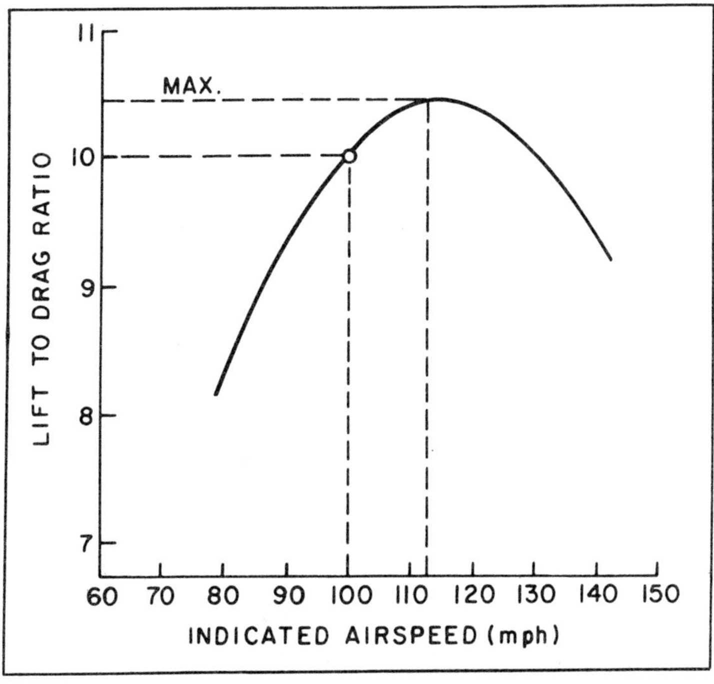

Fig. 8-9. Determining maximum L/D ratio from flight test data.

5. The glide distance can now be determined for any altitude. For 10,000 feet the distance is determined as:

$$d_g = \frac{(L/D)_{max} h}{5280} = \frac{10.4(10{,}000)}{5280} = 19.7 \text{ miles}$$

Landing Distance Example

Test Conditions:

- ☐ Average ground roll = 600 ft.
- ☐ Headwind = 5 kts.
- ☐ Pressure altitude = 2000 ft.
- ☐ Temperature = 38°F.
- ☐ Weight = standard weight.
- ☐ Stall speed = 55 kts.

Assume that the landing is to be made at full stall.

1. First, correct the ground roll to zero wind conditions.

$$d_o = d_w \left[\frac{V_L}{V_L - V_w}\right]^2 = 600 \left[\frac{55}{55 - 5}\right]^2 = 726 \text{ ft.}$$

2. Next, correct this value to standard sea level distance. From Figs. 1-3 and 1-4, a pressure altitude of 2000 ft. and temperature of 38° yield a value of ρ/ρ_o of 0.97.

$$d_{SSL} = d_o \left[\frac{\rho}{\rho_o}\right] = 726(.97) = 704 \text{ ft.}$$

This is the ground roll distance at standard sea level, now referred to as d_G.

3. Now calculate the air distance to clear a 50 foot obstacle. For full stall landing, the average approach speed is 1.15 V_s.

$$1.15\ V_s = 1.15(55) = 63 \text{ kts}$$

From a plot of L/D versus velocity in landing configuration, assume that we have obtained an L/D of 6.9 for V = 63 kts. From the equation for air distance for full stall landing and stall speed in knots, we can now obtain the air distance.

$$d_A = \left(\frac{L}{D}\right)\left(50 + 0.0305\ V_s^2\right)$$

$$= (6.9)\ [50 + 0.0305\ (55)^2] = 981 \text{ feet}$$

4. The total landing distance to clear a 50 foot obstacle is now:

$$d = d_G + d_A = 704 + 981 = 1685 \text{ feet}$$

Appendix A

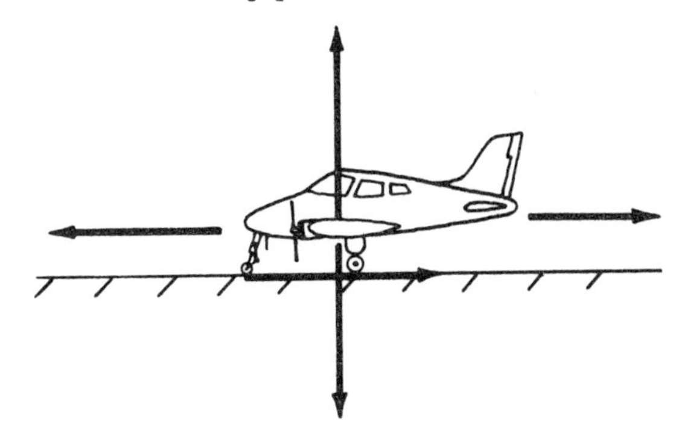

Symbols, Conversion Factors, Data Sheet

List of Symbols

bhp	brake horsepower
CAS	calibrated airspeed
D	drag
d	distance
d_A	air distance (takeoff or landing)
d_G	ground distance (takeoff or landing)
d_g	glide distance
d_o	distance with zero wind
d_{SSL}	distance at standard sea level
E	endurance
EGT	exhaust gas temperature
FC	fuel consumption
GR	glide ratio
GS	groundspeed
hp	horsepower
h	altitude
IAS	indicated airspeed
K	general constant
L	lift
L/D	lift to drag ratio
MAP	manifold air pressure
P	pressure; also, power
R	range
RC	rate-of-climb

rpm	revolutions per minute
RS	rate-of-sink
S	wing area
T	temperature; also, thrust
t	time
TAS	true airspeed
thp	thrust horsepower
V	velocity
V_c	cruise velocity
V_L	landing velocity
V_{to}	takeoff velocity
V_s	stall velocity
V_x	best angle-of-climb velocity
V_y	best rate-of-climb velocity
W	weight
α	angle of attack
ρ	density
ρ_o	standard sea level density

Conversion Factors

Multiply	By	To Get
Miles per hour (mph)	0.8696	Knots
Knots (kts)	1.1508	Miles per hour
Miles per hour (mph)	1.4667	Feet per second
Knots (kts)	1.6879	Feet per second
Miles (U.S.)	0.8696	Nautical miles
Nautical miles (nm)	1.1508	Miles (U.S.)
Miles	5280	Feet
Pounds/cu. inch	2.0360	Inches of Hg.
Inches of Hg.	0.4912	Pounds/cu. inch

Temperature:

$$°C = (°F - 32)\left(\frac{5}{9}\right)$$

$$°F = \left(\frac{9}{5}\right)(°C) + 32$$

Metric Conversion Factors

Multiply	By	To Get
Feet	0.3048	Meters
Gallons	3.7853	Liters
Horsepower	0.7457	Kilowatts
Knots	0.5151	Meters per second

Multiply	By	To Get
Miles per hour	0.4471	Meters per second
Pounds	4.448	Newtons
Pounds per sq. inch	6895	Pascals

Sample Data Sheet

FLIGHT TEST DATA

TEST: ________________

DATE: ________________

LOCATION: ________________

SL PRESS: ________________

SURFACE TEMP: ________________

PILOT: ________________

OBSERVERS: ________________

____

____

FUEL ________

WEIGHT

EMPTY ________

FUEL ________

PASS. ________

BALLAST ________

TOTAL T.O ________

LANDING ________

AVERAGE ________

RUN	IAS	P.ALT.	OAT	TIME	POWER		

REMARKS:

Appendix B

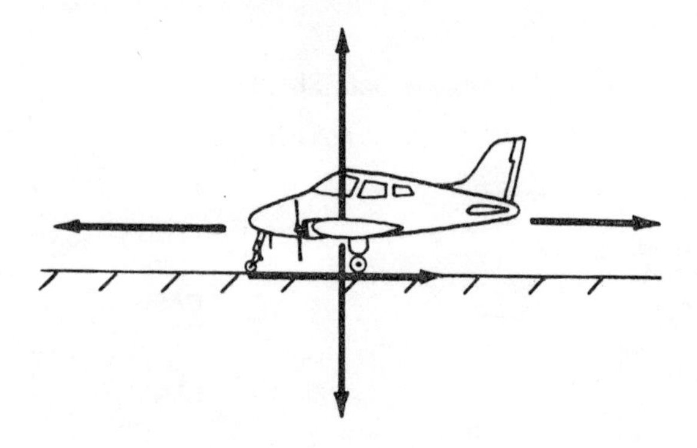

Performance Data Examples

This appendix contains actual performance data as presented in Pilot's Operating Handbooks. Stall, takeoff, climb, cruise, descent, and landing performance charts are included for the Piper Warrior II. The charts are typical of performance information presented in graph form. The vertical scale in most of these graphs is density altitude. Notice that instead of simply labeling the scale in density altitude, a chart for determining density altitude from pressure altitude and temperature is provided along the left hand side of each graph. The intersections of the standard temperature line and the various pressure altitudes would yield respective values of density altitude.

Performance in tabular form is shown for the Cessna 150. These tables also contain performance information on takeoff, climb, cruise, and landing. Interpolation is required for values between figures listed and the notes must be referred to for adjustment to special conditions.

All of this information is provided for illustrative use only. None of it should be used for the actual operation of the airplanes involved, since it is not kept current by appropriate revisions. *Neither the aircraft manufacturers (Piper Aircraft Corporation and Cessna Aircraft Company) nor the author and publisher will be responsible for the accuracy of the data in this appendix.*

This information is provided through the courtesy of Piper Aircraft Corporation and Cessna Aircraft Company.

PA-28-161

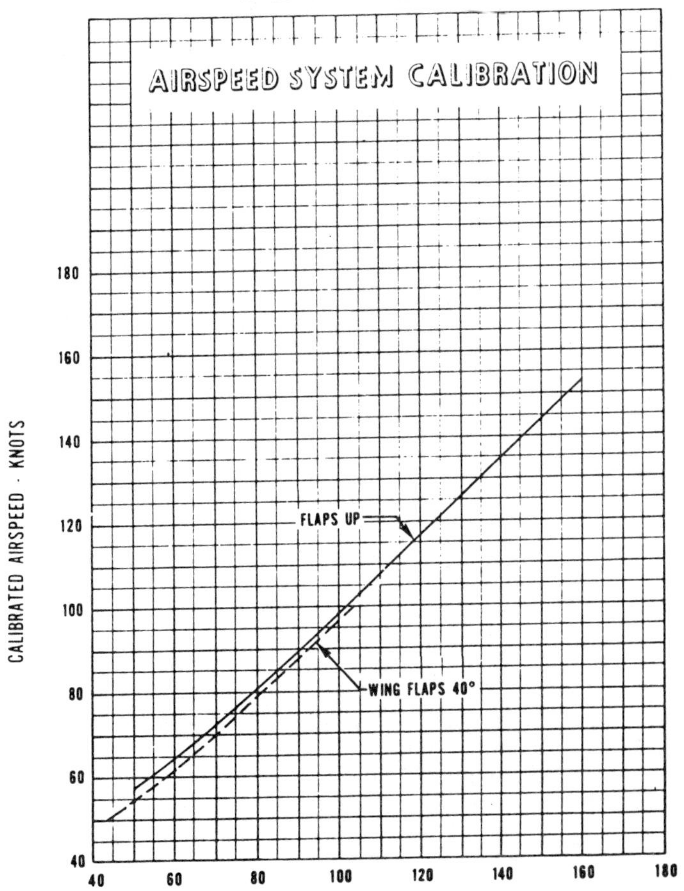

AIRSPEED SYSTEM CALIBRATION
CALIBRATED AIRSPEED - KNOTS
180
160
140
120
100
80
60
40
FLAPS UP
WING FLAPS 40°
40
60
80
100
120
140
160
180
INDICATED AIRSPEED - KNOTS

PA-28-161

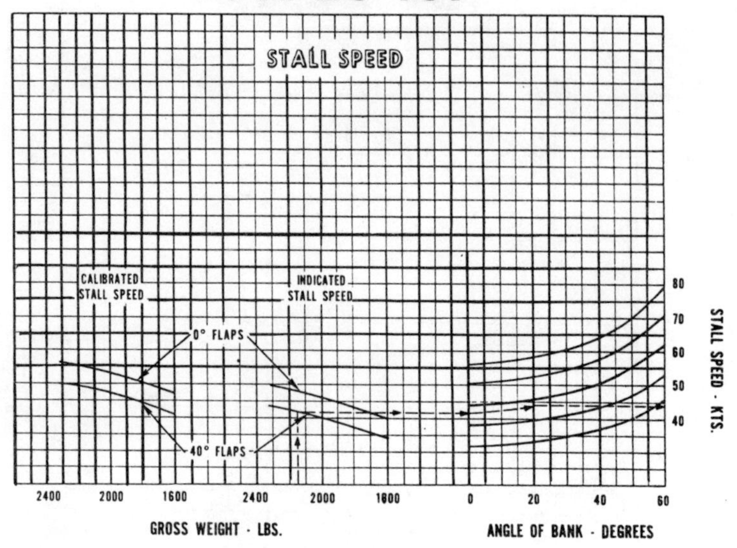

Example:
Gross weight: 2170 lbs.
Angle of bank: 20°
Flap position: 40°
Stall speed, indicated: 44 KTS

PA-28-161

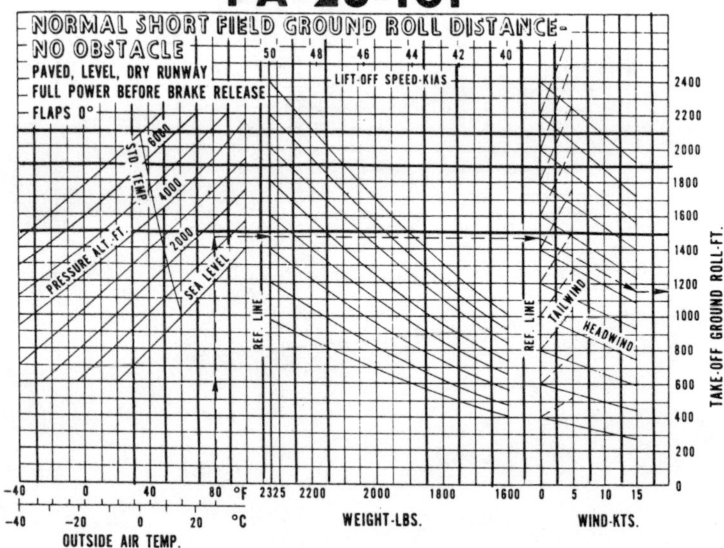

Example:
Departure airport pressure altitude: 1500 ft.
Departure airport temperature: 80° F
Weight: 2325 lbs.
Wind: 15 KTS headwind
Ground roll: 1150 ft.
Lift-off speed: 50 KIAS

PA-28-161

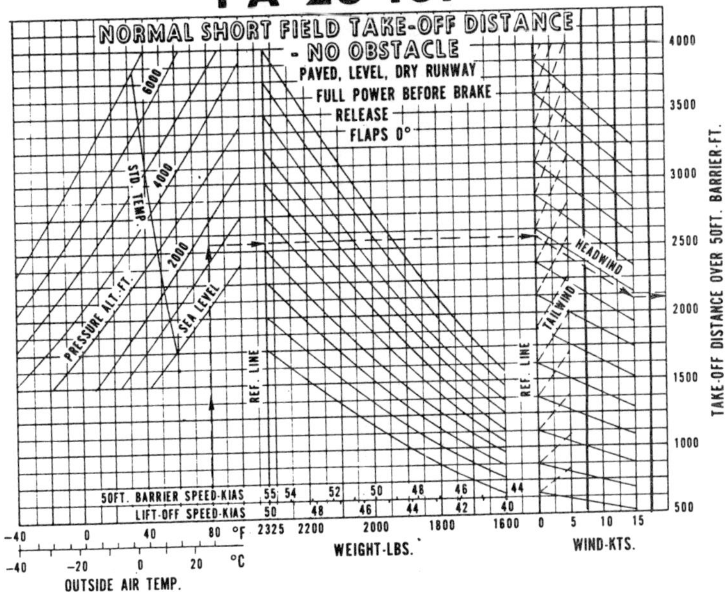

Example:
Departure airport pressure altitude: 1500 ft.
Departure airport temperature: 80°F
Weight: 2325 lbs.
Wind: 15 KTS headwind
Distance over 50 ft. barrier: 2100 ft.
Lift-off speed: 50 KIAS
Barrier speed: 55 KIAS

PA-28-161

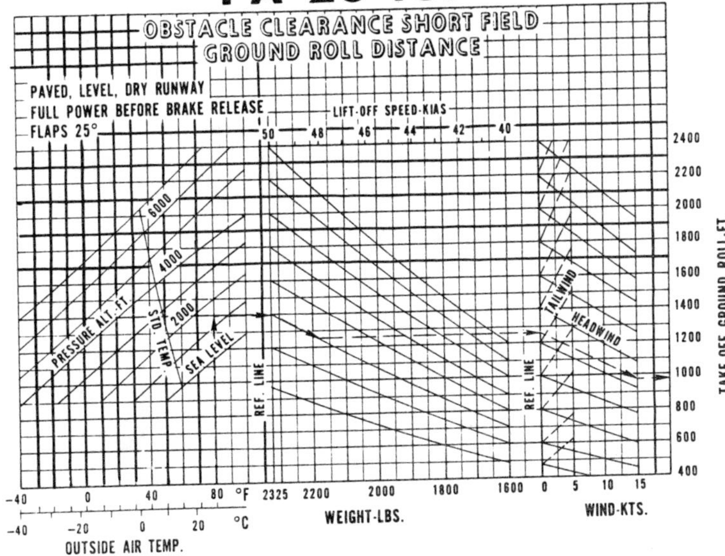

Example:
Departure airport pressure altitude: 1500 ft.
Departure airport temperature: 80°F
Weight: 2175 lbs.
Wind: 15 KTS headwind
Ground roll: 975 ft.
Lift-off speed: 48 KIAS

PA-28-161

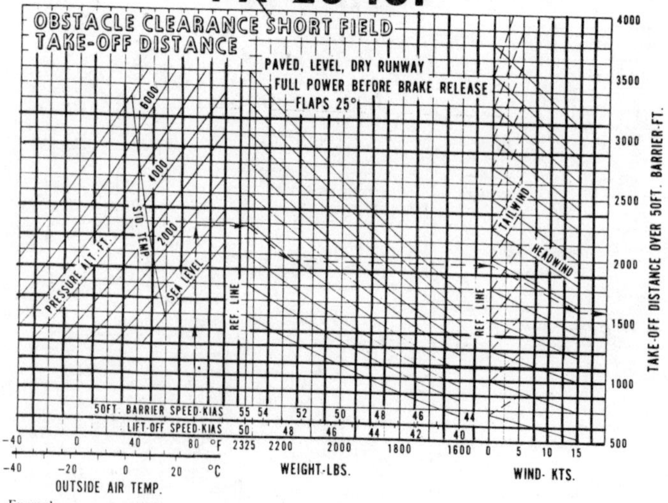

Example:

Departure airport pressure altitude: 1500 ft.
Departure airport temperature: 80° F
Weight: 2175 lbs.
Wind: 15 KTS headwind
Distance over 50 ft. barrier: 1600 ft.
Lift-off speed: 48 KIAS
Barrier speed: 53 KIAS

PA-28-161

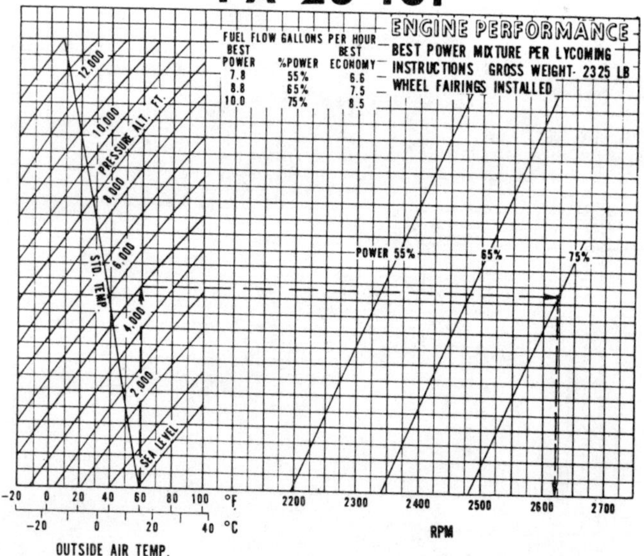

Example:

Cruise pressure altitude: 5000 ft.
Cruise OAT: 60° F
Cruise power: 75%
Engine RPM: 2620

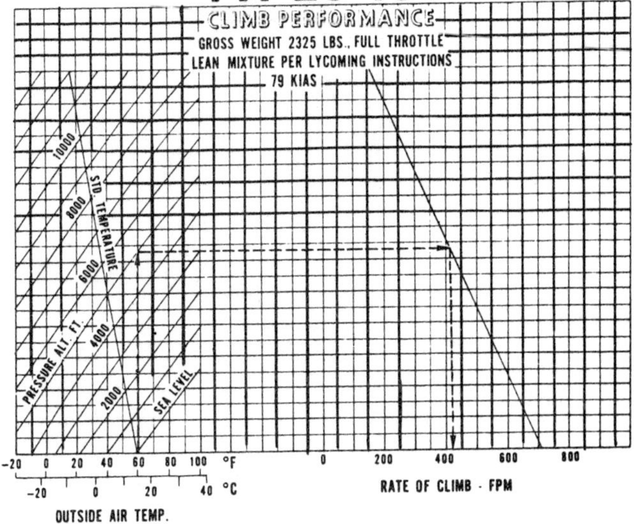

Example:
Climb pressure altitude: 5000 ft.
Climb OAT: 60°F
Rate of climb: 420 ft/min.

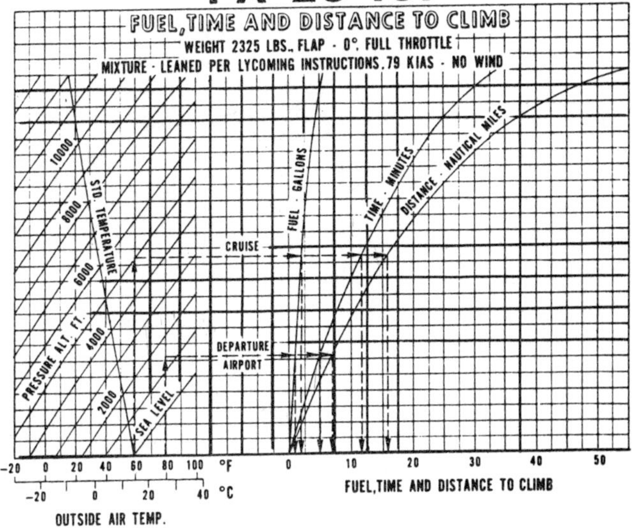

Example:
Departure airport pressure altitude: 1500 ft.
Departure airport temperature: 80°F
Cruise pressure altitude: 5000 ft.
Cruise OAT: 60°F
Time to climb (11.5 min. minus 5 min.): 6.5 min.
Distance to climb (15.7 miles minus 7 miles): 8.7 nautical miles
Fuel to climb (2 gal. minus 1 gal.): 1 gal.

PA-28-161

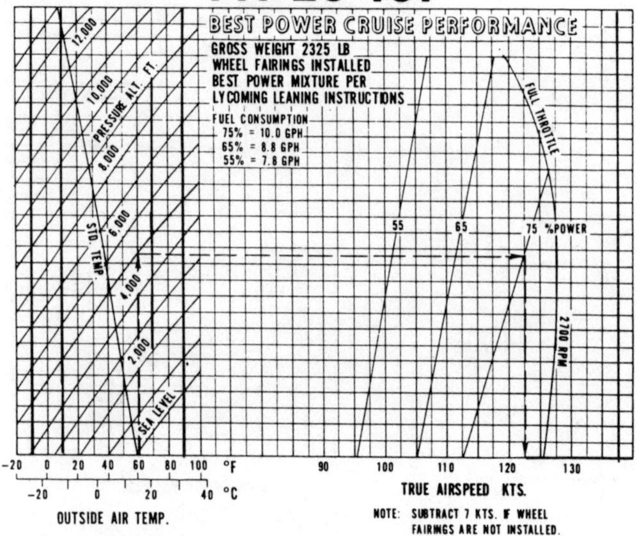

Example:

Cruise pressure altitude: 5000 ft.
Cruise OAT: 60°F
Cruise power: 75% best power mixture
Cruise speed: 122.5 KTS TAS

PA-28-161

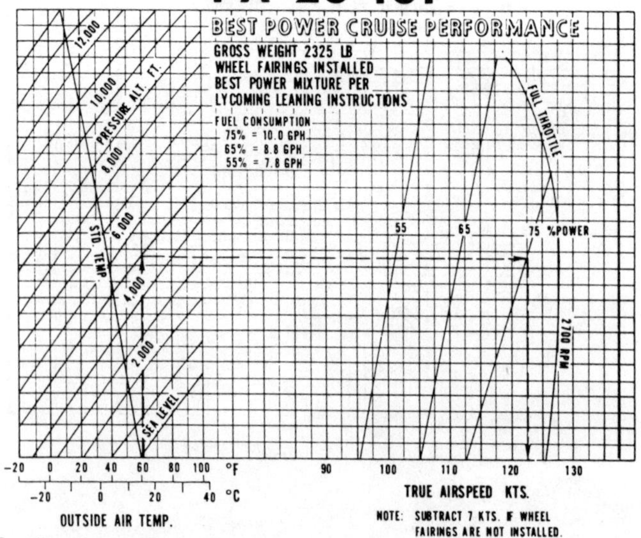

Example:

Cruise pressure altitude: 5000 ft.
Cruise OAT: 60°F
Cruise power: 75% best power mixture
Cruise speed: 122.5 KTS TAS

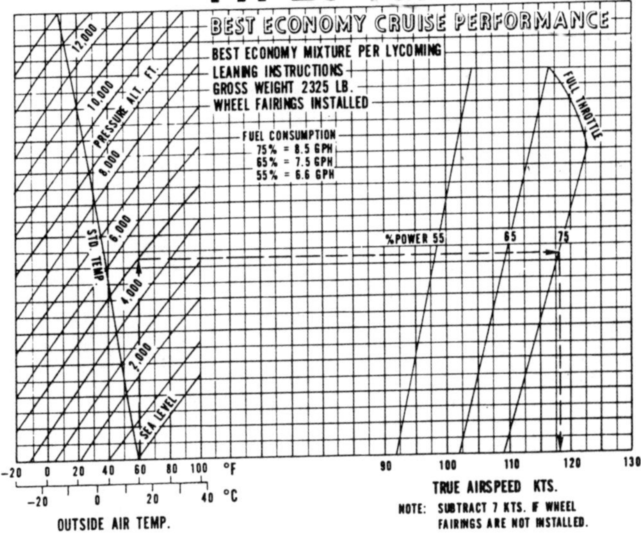

Example:
Cruise pressure altitude: 5000 ft.
Cruise OAT: 60°F
Cruise power: 75% best power mixture
Cruise speed: 118 KTS TAS

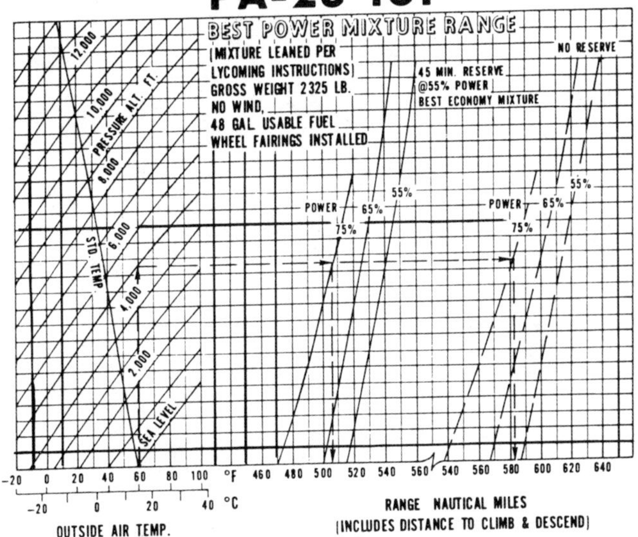

Example:
Cruise pressure altitude: 5000 ft.
Cruise OAT: 60°F
Cruise power: 75% best power mixture
Range w/45 min. reserve @ 55% power: 505 nautical miles
Range w/no reserve: 582 nautical miles

PA-28-161

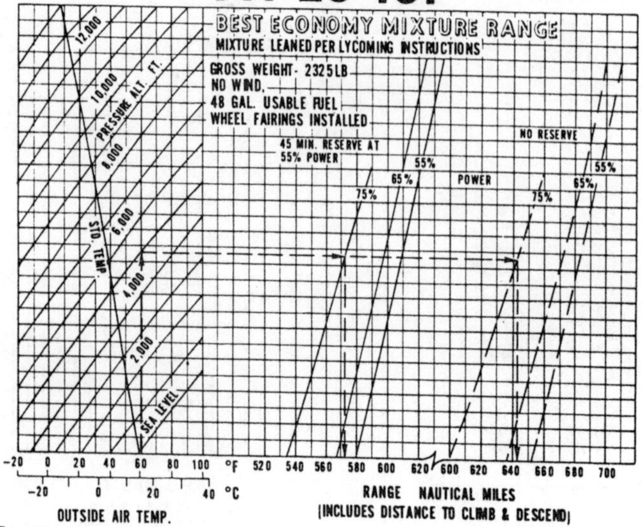

Example:
Cruise pressure altitude: 5000 ft.
Cruise OAT: 60°F
Cruise power: 75% best economy mixture
Range w/45 min. reserve @ 55% power: 572 nautical miles
Range w/no reserve: 642 nautical miles

PA-28-161

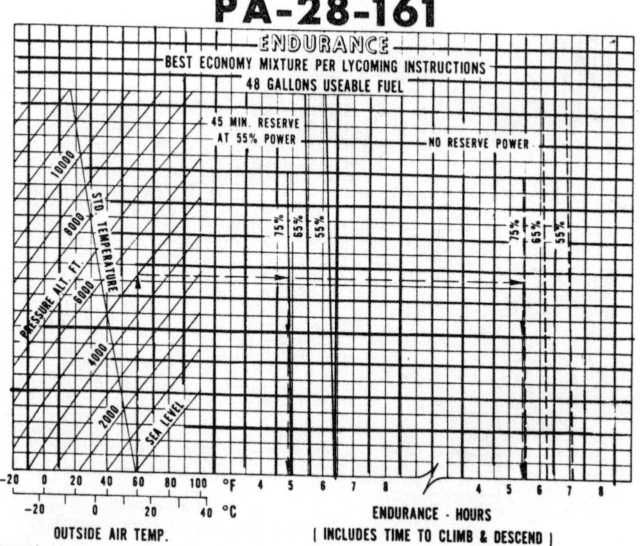

Example:
Cruise pressure altitude: 5000 ft.
Cruise OAT: 60°F
Cruise power: 75% best economy mixture
Endurance w/45 min. reserve @ 55% power: 4.85 hrs.
Endurance w/no reserve: 5.45 hrs.

PA-28-161

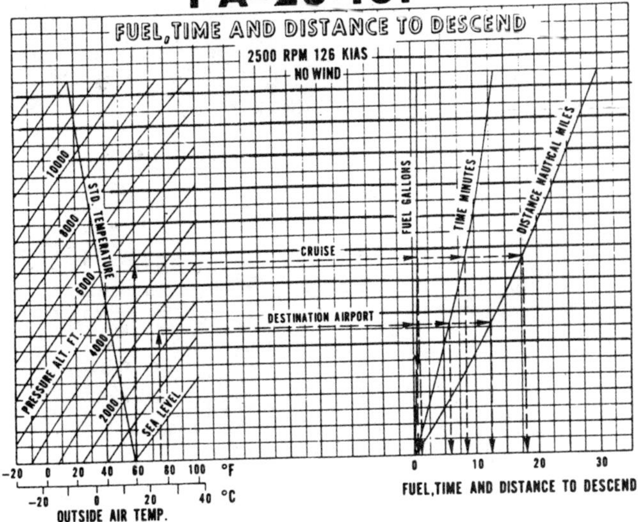

Example:

Destination airport pressure altitude: 2500 ft.
Destination airport temperature: 75°F
Cruise pressure altitude: 5000 ft.
Cruise OAT: 60°F
Time to descend (8.5 min. minus 6 min.): 2.5 min.
Distance to descend (18 miles minus 12.5 miles): 55 nautical miles
Fuel to descend: (1 gal. minus .5 gal.): .5 gal.

PA-28-161

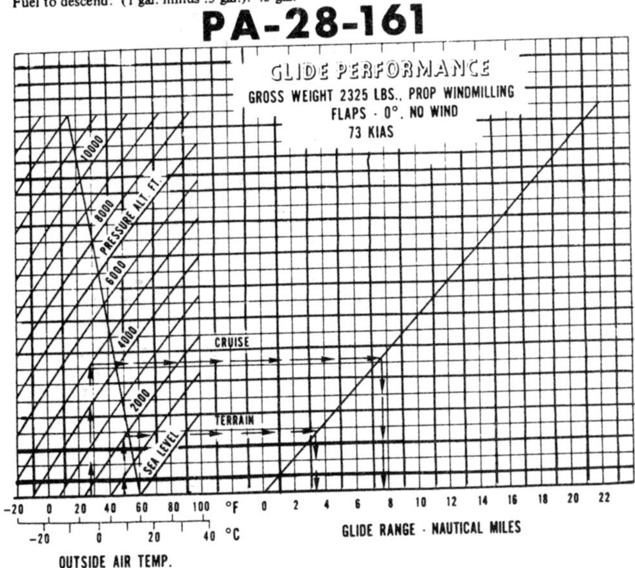

Example:

Cruise pressure altitude: 5000 ft.
Cruise OAT: 28°F
Terrain pressure altitude: 2000 ft.
Temperature at terrain: 50°F
Glide distance (7.7 miles minus 3.5 miles): 4.2 nautical miles

PA-28-161

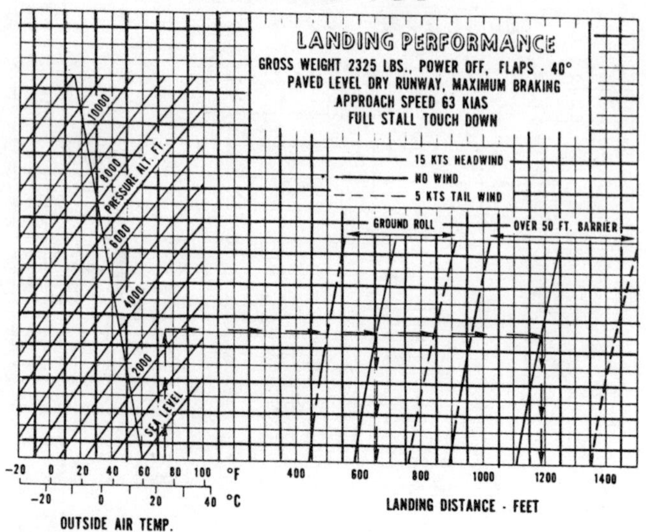

Example:

Destination airport pressure altitude: 2500 ft.
Destination airport temperature: 75°F
Destination airport wind: 0 KTS
Ground roll: 660 ft.
Distance over 50 ft. barrier: 1190 ft.

STALL SPEEDS

CONDITIONS:
Power Off

NOTES:
1. Altitude loss during a stall recovery may be as much as 160 feet.
2. KIAS values are approximate and are based on airspeed calibration data with power off.

MOST REARWARD CENTER OF GRAVITY

WEIGHT LBS	FLAP DEFLECTION	ANGLE OF BANK							
		0°		30°		45°		60°	
		KIAS	KCAS	KIAS	KCAS	KIAS	KCAS	KIAS	KCAS
1670	UP	36	46	39	49	43	55	51	65
	10°	36	43	39	46	43	51	51	61
	30°	31	41	33	44	37	49	44	58

MOST FORWARD CENTER OF GRAVITY

WEIGHT LBS	FLAP DEFLECTION	ANGLE OF BANK							
		0°		30°		45°		60°	
		KIAS	KCAS	KIAS	KCAS	KIAS	KCAS	KIAS	KCAS
1670	UP	40	48	43	52	48	57	57	68
	10°	40	46	43	49	48	55	57	65
	30°	35	43	38	46	42	51	49	61

TIME, FUEL, AND DISTANCE TO CLIMB

CONDITIONS:
Flaps Up
Full Throttle
Standard Temperature

MAXIMUM RATE OF CLIMB

NOTES:
1. Add 0.8 of a gallon of fuel for engine start, taxi and takeoff allowance.
2. Mixture leaned above 3000 feet for maximum RPM.
3. Increase time, fuel and distance by 10% for each 10°C above standard temperature.
4. Distances shown are based on zero wind.

WEIGHT LBS	PRESSURE ALTITUDE FT	TEMP °C	CLIMB SPEED KIAS	RATE OF CLIMB FPM	FROM SEA LEVEL		
					TIME MIN	FUEL USED GALLONS	DISTANCE NM
1670	S.L.	15	67	715	0	0	0
	1000	13	66	675	1	0.2	2
	2000	11	66	630	3	0.4	3
	3000	9	65	590	5	0.7	5
	4000	7	65	550	6	0.9	7
	5000	5	64	505	8	1.2	9
	6000	3	63	465	10	1.4	12
	7000	1	63	425	13	1.7	14
	8000	-1	62	380	15	2.0	17
	9000	-3	62	340	18	2.3	21
	10,000	-5	61	300	21	2.6	25
	11,000	-7	61	255	25	3.0	29
	12,000	-9	60	215	29	3.4	34

CONDITIONS:
1670 Pounds
Recommended Lean Mixture

CRUISE PERFORMANCE

CESSNA
MODEL 152

NOTE:
Cruise speeds are shown for an airplane equipped with speed fairings which increase the speeds by approximately two knots.

PRESSURE ALTITUDE FT	RPM	20°C BELOW STANDARD TEMP			STANDARD TEMPERATURE			20°C ABOVE STANDARD TEMP		
		% BHP	KTAS	GPH	% BHP	KTAS	GPH	% BHP	KTAS	GPH
2000	2400	---	---	---	75	101	6.1	70	101	5.7
	2300	71	97	5.7	66	96	5.4	63	95	5.1
	2200	62	92	5.1	59	91	4.8	56	90	4.6
	2100	55	87	4.5	53	86	4.3	51	85	4.2
	2000	49	81	4.1	47	80	3.9	46	79	3.8
4000	2450	---	---	---	75	103	6.1	70	102	5.7
	2400	76	102	6.1	71	101	5.7	67	100	5.4
	2300	67	96	5.4	63	95	5.1	60	95	4.9
	2200	60	91	4.8	56	90	4.6	54	89	4.4
	2100	53	86	4.4	51	85	4.2	49	84	4.0
	2000	48	81	3.9	46	80	3.8	45	78	3.7
6000	2500	---	---	---	75	105	6.1	71	104	5.7
	2400	72	101	5.8	67	100	5.4	64	99	5.2
	2300	64	96	5.2	60	95	4.9	57	94	4.7
	2200	57	90	4.6	54	89	4.4	52	88	4.3
	2100	51	85	4.2	49	84	4.0	48	83	3.9
	2000	46	80	3.8	45	79	3.7	44	77	3.6
8000	2550	---	---	---	75	107	6.1	71	106	5.7
	2500	76	105	6.2	71	104	5.8	67	103	5.4
	2400	68	100	5.5	64	99	5.2	61	98	4.9
	2300	61	95	5.0	58	94	4.7	55	93	4.5
	2200	55	90	4.5	52	89	4.3	51	87	4.2
	2100	49	84	4.1	48	83	3.9	46	82	3.8
10,000	2500	72	105	5.8	68	103	5.5	64	103	5.2
	2400	65	99	5.3	61	98	5.0	58	97	4.8
	2300	58	94	4.7	56	93	4.5	53	92	4.4
	2200	53	89	4.3	51	88	4.2	49	86	4.0
	2100	48	83	4.0	46	82	3.9	45	81	3.8
12,000	2450	65	101	5.3	62	100	5.0	59	99	4.8
	2400	62	99	5.0	59	97	4.8	56	96	4.6
	2300	56	93	4.6	54	92	4.4	52	91	4.3
	2200	51	88	4.2	49	87	4.1	48	85	4.0
	2100	47	82	3.9	45	81	3.8	44	79	3.7

CESSNA
MODEL 152

TAKEOFF DISTANCE

SHORT FIELD

CONDITIONS:
Flaps 10°
Full Throttle Prior to Brake Release
Paved, Level, Dry Runway
Zero Wind

NOTES:
1. Short field technique as specified in Section 4.
2. Prior to takeoff from fields above 3000 feet elevation, the mixture should be leaned to give maximum RPM in a full throttle, static runup.
3. Decrease distances 10% for each 9 knots headwind. For operation with tailwinds up to 10 knots, increase distances by 10% for each 2 knots.
4. For operation on a dry, grass runway, increase distances by 15% of the "ground roll" figure.

WEIGHT LBS	TAKEOFF SPEED KIAS		PRESS ALT FT	0°C		10°C		20°C		30°C		40°C	
	LIFT OFF	AT 50 FT		GRND ROLL	TOTAL TO CLEAR 50 FT OBS	GRND ROLL	TOTAL TO CLEAR 50 FT OBS	GRND ROLL	TOTAL TO CLEAR 50 FT OBS	GRND ROLL	TOTAL TO CLEAR 50 FT OBS	GRND ROLL	TOTAL TO CLEAR 50 FT OBS
1670	50	54	S.L.	640	1190	695	1290	755	1390	810	1495	875	1605
			1000	705	1310	765	1420	825	1530	890	1645	960	1770
			2000	775	1445	840	1565	910	1690	980	1820	1055	1960
			3000	855	1600	925	1730	1000	1870	1080	2020	1165	2185
			4000	940	1775	1020	1920	1100	2080	1190	2250	1285	2440
			5000	1040	1970	1125	2140	1215	2320	1315	2525	1420	2750
			6000	1145	2200	1245	2395	1345	2610	1455	2855	1570	3125
			7000	1270	2470	1375	2705	1490	2960	1615	3255	1745	3590
			8000	1405	2800	1525	3080	1655	3395	1795	3765	1940	4195

CESSNA
MODEL 152

LANDING DISTANCE

SHORT FIELD

CONDITIONS:
Flaps 30°
Power OFF
Maximum Braking
Paved, Level, Dry Runway
Zero Wind

NOTES:

1. Short field technique as specified in Section 4.
2. Decrease distances 10% for each 9 knots headwind. For operation with tailwinds up to 10 knots, increase distances by 10% for each 2 knots.
3. For operation on a dry, grass runway, increase distances by 45% of the "ground roll" figure.

WEIGHT LBS	SPEED AT 50 FT KIAS	PRESS ALT FT	0°C		10°C		20°C		30°C		40°C	
			GRND ROLL	TOTAL TO CLEAR 50 FT OBS	GRND ROLL	TOTAL TO CLEAR 50 FT OBS	GRND ROLL	TOTAL TO CLEAR 50 FT OBS	GRND ROLL	TOTAL TO CLEAR 50 FT OBS	BRIND ROLL	TOTAL TO CLEAR 50 FT OBS
1670	54	S.L.	450	1160	465	1185	485	1215	500	1240	515	1265
		1000	465	1185	485	1215	500	1240	520	1270	535	1295
		2000	485	1215	500	1240	520	1270	535	1300	555	1330
		3000	500	1240	520	1275	540	1305	560	1335	575	1360
		4000	520	1275	540	1305	560	1335	580	1370	600	1400
		5000	540	1305	560	1335	580	1370	600	1400	620	1435
		6000	560	1340	580	1370	605	1410	625	1440	645	1475
		7000	585	1375	605	1410	625	1440	650	1480	670	1515
		8000	605	1410	630	1450	650	1480	675	1520	695	1555

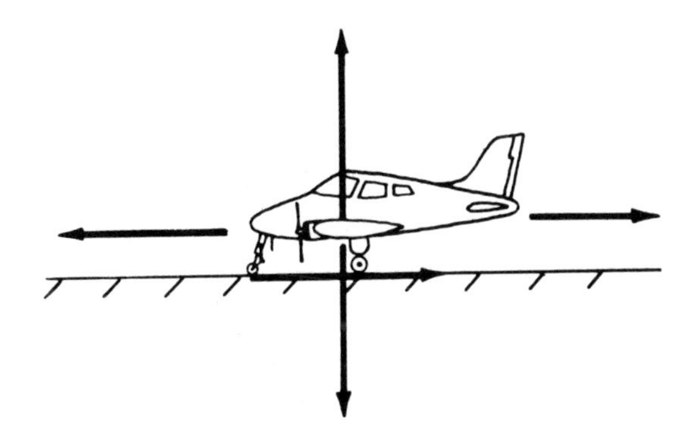

Index

W